The Dynamics of a Coalfield

Land, Space and Resource Exploitation During the Industrial Revolution

By

William Hartley C.Geog. FRGS

Published by New Generation Publishing in 2014

Copyright © William Hartley 2014

First Edition

The author asserts the moral right under the Copyright, Designs and Patents Act 1988 to be identified as the author of this work.

All Rights reserved. No part of this publication may be reproduced, stored in a retrieval system or transmitted, in any form or by any means without the prior consent of the author, nor be otherwise circulated in any form of binding or cover other than that in which it is published and without a similar condition being imposed on the subsequent purchaser.
www.newgeneration-publishing.com

 New Generation **Publishing**

Introduction

Our civilisation as George Orwell[1] remarked is built on coal. Despite this there have been only two empirical studies of the British coal industry. Otherwise the regional study describing activity in different parts of the British coalfield has been left to fill the gap. It is curious that this should be the case since without coal none of the industries apparently of greater interest to the historian; railways, the textiles trades and shipbuilding for example, would have existed in the same scale and form. Coal mining was not however simply an extractive industry whose product measured in millions of tons provided a footnote for the historian more interested in the trades it powered. Behind the industry lay the complex interplay of land use competition. Coal mining is a messy and destructive industry, vital of course but in many ways a nuisance for the landowner wishing to gain revenue from a variety of sources, some of which might be adversely affected by the presence of a colliery.

If the question is asked; why did the industrial revolution occur in Britain rather than continental Europe? The answer lies in three words; Coal, Climate and Space. Like Britain the countries of Northern Europe possessed both coal reserves and a cool temperate climate. Unlike Britain what they also had in abundance was space. The primary fuel source of the pre industrial age was timber, which is also an important building material. In Britain the amount of accessible timber was less in the sixteenth century than today. Importing timber in the small ships of the pre industrial era would have been difficult and expensive. This meant a much earlier recourse to a secondary fuel source, coal. In continental Europe with its

---

[1] Orwell G. *Essays*, (London 1946)

large navigable rivers and huge forests, access to timber supplies from distant sources was much easier; the limited space in an island where the timber reserves were more rapidly denuded made coal mining a necessity and this occurred at an earlier date than had previously been thought to be the case. The long held assumption that mining technology was brought into Britain from the German silver mines during the fourteenth century has been shown to be false. Recent archaeological evidence from the North East of England has revealed that sophisticated mining techniques were in use at least a hundred years earlier. These techniques closely linked to agriculture are an important key to why the coal mining industry developed in the way that it did in the district which forms the subject of this study.

In some parts of Britain the coal industry dominated to such an extent that everything else seemed subordinate to its needs. Housing, transport, the landscape itself, was shaped and as necessary distorted by the need to win coal. This created a legacy visible well into the twentieth century; for example the communities which had to adapt to life in the shadow of the workplace: quite literally where mining produced giant spoil heaps as evidence of what was going on underground. Even for those who never went there, the major coal mining regions of Britain were instantly recognisable.

Elsewhere though, where mining was not the dominant enterprise coal had to fight for its place. What began as a necessity powering the creation of other industries in the locality, eventually turned into a marginal activity as supplies from distant coalfields began to be brought in by rail. Paradoxically it is in such areas, where coal mining was not the dominant industrial activity that the whole complex picture of land use competition and the fight for space during the industrial revolution can be revealed. The more marginal the coalfield the less likely that much in the way of physical evidence will have survived. Indeed it is quite possible that many modern inhabitants of such areas are unaware that a mining industry ever existed. For example the arrival of a new railway line could in a matter of weeks bring about the demise of a coalfield. The West Riding of Yorkshire contained such a coalfield and it will be used to illustrate the themes of this book: land use, the competition for space and the decision making process in a time of industrial and urban expansion. Although the industry will inevitably be examined in some detail it is not intended to be a further addition to the many regional studies which have been published. Rather it is an exploration of ideas, testing theories against what was actually happening on the ground.

This coalfield provides all the key features. It sprang from agriculture, being at one time a by product of land from which any source of income had to be exploited out of necessity. The landscape was often hilly with isolated pockets of population. There was a brief period when the locational advantages of a local supply made mining important and allowed the expansion of other industries. Then came the railways which ought to have led to the demise of an industry comprising mostly small units. Paradoxically although marginalised by the availability of supplies from beyond the District delivered by modern means of transportation, output increased. Finally and well before the Second World War the last traces disappeared placing it beyond human memory.

Although the landscape of the Huddersfield, Halifax and Bradford is dominated by the wool textiles industry, this only came about because local coal was available to facilitate the expansion. Before the advent of steam power, woollen mills were crowded along valley bottoms gaining their energy needs from rivers. Close to the mills settlements once given over to domestic weaving began to expand. Villages grew into towns, which in turn absorbed their neighbours, turning them into suburbs. Within a small area constrained by steep topography, land use competition became intense. As the use of steam began to liberate mills from valley bottom sites so they grew into huge industrial barracks, requiring a great deal of space.

The workforce necessary for these enterprises needed to be housed, usually in rows of terraced properties again to maximise land use. In addition there were the supporting industries, together with roads and railways needed to create the framework of the industrial revolution.

Amidst this frantic expansion which absorbed so much of the nineteenth century, the coal industry of the District had to find its place. Although coal was ubiquitous the most intensive activity took place within the boundaries of the former county borough of Huddersfield, along a narrow corridor of land which ran across country to the east of Halifax, then northwards ending on the southern outskirts of Bradford. Here the pattern of landownership consisted of several estates owned by members of the aristocracy and gentry. Most of this land had been acquired after the dissolution of the monasteries. Like the great monastic houses the new owners had originally used this land for sheep farming, which continued to be the case until the fast flowing streams and their soft water made the district an obvious location for the transition of cloth weaving from a domestic into an industrial activity.

This book will attempt to examine the development of the local coal mining industry in Huddersfield, Halifax and part of Bradford, not as a standard history but rather to analyse land use competition, resource exploitation and the way that landowners shaped the development of the coal industry through the policies they adopted. The book is designed to provide an analytical process that shows how geography can shape economic activity. Throughout where the term 'District' is used it refers to the map shown in fig.1. Where the term 'small colliery' is used it denotes a colliery producing less than 10,000 tons of coal per annum. Above this figure a greater degree of mechanisation was likely. Below this level of production the mining operation was likely to be extremely primitive. A 'coal master' describes an individual directly involved in the mining of coal rather than a landowner leasing mineral rights to another party.

The geography of economic growth throws up both regional and inter regional differences. Finding a closed system for the purpose of study and analysis is likely to be impossible. This is particularly the case in Britain where even in the early stages of the industrial revolution the manufacturing districts were already starting to become crowded and complicated places. The area of the West Riding which forms the basis for this study does however present features which come near to a closed system, though of course this was to erode over time. The significance of topography played a part in this. The steep sided valleys provided natural barriers to effective transportation of a heavy bulk commodity. Although canals had penetrated to some extent, their impact was not felt in the more remote areas, meaning that a high degree of sub regional autonomy existed until the 1840s when railway penetration gathered pace. The initial loci of economic growth were the valley bottom locations but as noted these had become crowded by the early years of the nineteenth century and movement beyond was only possible because of the existence of local coal supplies. Economic expansion has a spatial as well as a sectorial dimension but the spatial element cannot be explained purely by the locational advantage of a local coal supply.

Economic location theory is founded on the assumption that economic activities in general and locational choice in particular are governed by the desire to make nominal profits. Differences between money receipts and money costs means that the size of this difference will vary through space as the costs of operation and receipts vary. However the viability of a colliery in a marginal mining area will be affected by factors beyond merely the location and the market. The 'Least Cost' approach of Alfred Weber[2] is predicated on the assumption that markets exist at a particular point where demand is constant or unlimited: the cost of resources required in production,

---

[2] Weber A. *Theory of the Location of Industries* (Chicago 1929)

prices at source, the cost of hauling the unit of finished goods to the market, meaning that the point of minimum cost may be determined. However this theory cannot accommodate variations in demand.

The market area theory of August Lösch[3] expresses the influence of variation in prices and demand. Unlike Weber's theory it encompasses all the firms in an industry simultaneously so that optimum location patterns not just the optimum site for the next firm to enter an industry can be determined for an equilibrium situation.

Neither theory can accommodate spatially variable demand and costs, indeed no theory exists which will allow the ascertainment of the optimum location pattern of a coal industry since coal cannot be regarded as a ubiquity on a coalfield because geological factors prevent this happening. Further, the picture becomes even more complicated by such factors as coal types, the thickness of seams and viable areas of coal to lease.

The chief coal consuming industry in the District was the wool textiles industry but close on its heels came the many support trades, together with the needs of the expanding population and in turn the industries required to serve it. A colliery is an expensive undertaking. To a considerable extent the costs are quite literally sunk: little can be recovered should the mine fail. This means then that investors in a colliery enterprise will seek to reduce the risk of failure by developing the mine on property with proven reserves and access to markets. The risk cannot entirely be eliminated but if a coal with a good economic value and a reasonable seam thickness exists in the area, then a colliery stands a chance of succeeding. Although operating conditions in Bradford Dale were quite different to Huddersfield and Halifax, methods of working were broadly the same. This only began to change as the pace of railway penetration increased. Prior to the arrival of the railway there was a certain ubiquity based on the fact that

---

[3] Losch A. *The Economics of Location* (New York 1940)

coal was a heavy and difficult commodity to transport.

As a starting point then the Market Area Model is perhaps the best basis on which to work. This though only presents a convenient entry point into a complex problem. Market area is of course a means for explaining location within a given space. There are though what might be described as vertical factors affecting the explanation and indeed one which at first glance hardly seems a consideration in mining: *mobility*. In order to gain a better explanation than is likely to be yielded up purely by the application of the Market Area Model, the intention is to apply an extra layer of enquiry shaped by Game Theory[4]. The reason for attempting this is because of the landownership pattern. The coalfield of the District was not subjected to sophisticated mining techniques, save on the holdings of the Bradford iron companies and one other location. Mining sprang from agriculture and the approach might be more instinctive than technical. Few landowners directly interested themselves in mining. Their role was to balance the needs of this intrusive industry against competing demands for land and earn a good income. Market Area Theory can take us into this enquiry; Game Theory may better explain what was going on.

The use of such an approach is encouraged by the fact that in the district there were both landowners who took a close personal interest in the development of their mineral holdings and others who preferred to rely upon an agent. Beneath this tier were the leaseholders; small coal masters, sometimes referred to as 'pit takers'. These were men who used their own skill and judgement when sinking a mine either to exploit it for their own personal gain or because they had been contracted to do so by an industrialist. It means then that we add to the geological and topographical factors the additional variables of mineralogy (the fact that more than one mineral not just coal, might be exploited) to the policy and prejudices of

---

[4] Neumann J. & Morgenstern O. *Theory of Games and Economic Behaviour* (P.U.P). 2nd edit.1947).

the landowner and those working his leaseholds.

When the decision making process is added then a further analytical tool becomes helpful. Game Theory is a mathematical tool for the decision maker and the strength of this approach lies in its ability to structure and analyse problems of strategic choice. Game Theory is a formal model of an interactive situation which involves several players. The formal definition lays out the players, their preferences, their information and the strategic choices open to them. In the exploitation of coal reserves this might involve the location of the coalfield, access to markets, competition, the needs of consumers and the types of coal and other minerals available.

Cooperative Game Theory investigates such coalitional games with respect to the relative amounts of power held by various players or how a successful coalition should divide its proceeds. Non Cooperative Game Theory is concerned with the analysis of strategic choices. The paradigm of Non Cooperative Game Theory is that the details of the ordering and timing of players' choices are crucial to determining the outcome of a game. The latter model is then considered to be most appropriate for the investigation of the decision making process in land use and resource exploitation.

In order to test the theory it is first necessary to explain the significance of the various factors that might be brought into play. As a consequence chapters have been devoted to such subjects as the geology of the District, landownership, production and so forth in order to build a detailed picture of the dynamics at work. With this information assembled the final chapters will concentrate on The Game.

Chapter One

The Economic Geology

An important preliminary is to gain an understanding of the mineral wealth of the District. This includes not only the minerals found in association with coal but also the different types of coal. This needs to be understood before other factors such as the pattern of landownership are introduced and the variables affecting the decision making process are set out. What is referred to in the text as the 'District' encompasses most of the former county borough of Huddersfield, a narrow corridor of land connecting it to the east of the former county borough of Halifax and then through the small townships of Southowram and Northowram to the southern outskirts of the City of Bradford, sometimes referred to as Bradford Dale.

A significant part of the landholdings in the District were originally monastic properties. The great religious houses of Fountains, Rievaulx and Jervaulx all owned land in the area. This was not though the well ordered heavily timbered aspect of the typical English champaign[1]. Instead it was mostly marginal land farmed to some extent in the valley bottoms but left largely unimproved by the monastic houses. The chief value lay in the grazing on higher ground and it was this which made such property attractive to those who bought these lands from the Crown in the sixteenth century following the Reformation. Huge areas of waste and moorland provided the acreage for grazing sheep. For those living in remote valleys with access only to the more marginal land there was an obvious incentive to utilise any raw materials that might extracted from beneath the ground. The monastic houses had been involved in the granting of leases to work coal

---

[1] Knowles D. *The Religious Orders in England and Wales* 5th edit. (London 1974) p44

and stone, consequently mining and quarrying expertise was available in the District centuries before the industrial revolution and had been long established as a way of augmenting income from the land. This ability to undertake small scale mining and quarrying was to be the foundation of resource exploitation in the industrial age, providing as it did knowledge and techniques which enabled some industrialists to both fund as well as power their enterprises. Agriculture and mineral resource exploitation were closely allied and this continued to be the case well into the nineteenth century.

Legally estate owners were free to dig up whatever coal was found beneath their land without concern that the Crown was able to lay claim to it. In contrast this was not the case across the channel where European monarchs often claimed minerals found on private property. In England under the Forest Charter signed in 1217 just after Magna Carta, the Crown had already yielded to estate holders the ownership of both timber and any peat that was under their land and coal fell into the same category[2]. The ability to locate and then work outcrops of coal and stone meant that a trial and error approach to mining and other forms of mineral exploitation gradually evolved into mining skills that might be found among the populace of any village or hamlet in the District.

The dominance of an economy based on sheep farming persisted until the late eighteenth century. However with the growth of the northern industrial towns there was a change in land use. Turning over even marginal land from grazing to agriculture became economically viable as the urban market for foodstuffs grew. The field boundaries still to be found in the District illustrate this intake from moorland and waste to improve land for agricultural purposes. Even so amongst these new field boundaries opportunities to work minerals were not neglected. An outcrop of coal or the potential to create a simple box

---

[2] Freese B. *Coal A Human History*, (London 2005) p45

quarry to extract stone might intrude within the neat right angles of a newly enclosed field. It was a short lived phenomenon. Agriculture could never be sustained for long on the high contours of the valleys which ran down from the Pennines. Competition from more productive farming regions soon rendered such activity uneconomic. It was in fact the valley bottoms with their fast flowing streams providing the hard water so essential for the wool trade where the economic future lay. Cloth weaving previously a domestic activity was transferred to the mills which utilised the streams to provide the energy required to drive the looms. It is these mill buildings which shape the architectural legacy of the District. The evidence of resource exploitation is harder to find and yet without the expertise gained in the pre industrial era of extracting minerals for domestic use, it would have been much harder to develop what was to follow. Coal was only one of a range of minerals often occurring in association with each other that came to be utilised for industrial purposes; sandstone, iron ore, fireclay and ganister were all worked. Where they occurred in association this had the benefit of enhancing the value of a mineral which may have been uneconomic to work on its own.

The earliest references to coal mining in the District are to be found in legal documents dating back to the thirteenth century[5]. By the fifteenth and sixteenth centuries there are increasing mentions of coal and mineral rights, particularly in records of the court leets of the Manor of Wakefield[6]. This was perhaps the earliest evidence of coal being viewed as a commodity capable of enhancing the value of land. Further weight is given to this view by the way in which distinctions were beginning to be made between the different types of coal found in the District. One of the best known was the 'Soft Bed' which had the advantage of being found at shallow depths. This coal was relatively low in sulphur content making it useful for iron

---

[5] Walker J.W., *Wakefield Its History & People* (1939) p71
[6] Walker *Ibid*

working. Gradually a lexicon of coal types was created based on usage. Each was to find a niche in the local economy. The chief engine coal found in the district was the 'Hard Bed'. This was a type of coal with a high sulphur content because of the presence of pyrites. The Hard Bed coal provides an early example of a subsidiary industry being created as a consequence of coal mining. This coal was used in the manufacture of sulphuric acid. By 1840 there were three collieries in Halifax involved in mining coal which was used for this purpose[7].

One of several Parliamentary reports into the coal trade described the Huddersfield and Halifax coals as, 'gannister coals' adding that

'Soft' or coking coal is of 'good quality' in Halifax but 'poor' in Huddersfield. The report noted that seams varied in quality and thickness in each area with 'Soft' coal being between 1ft 6 ins and 3ft in thickness, lying at a depth of 110ft and 'Hard' being 90ft above the Soft, the coal being worked by driving levels to the boundary 50 yards apart. It noted that Hard Bed was around 2ft thick and that '1,250 tons per acre can be obtained'. The associated mineral ganister, was 1ft 2 ins in thickness and fireclay 3ft 4ins[8].

By the eighteenth century Bradford Dale was beginning to be seen as an attractive and competitive area for industry. Iron ore was the chief draw but with it were found coal types known locally as the 'Black Bed' and 'Better Bed'; the latter occurring in association with the iron ore and being described as a 'bituminous coal, dense, bright in colour and singularly free from sulphur and other impurities which unfit a coal for smelting purposes. It is chiefly used by the iron companies for the purpose of the furnace and the forge and it is the purity of the coal that the bars known as Low Moor, Bowling, and Farnley iron are known'[9].

---

[7] Stephens J.G., et al, *Geology of the Country Around Bradford* (1933) p77
[8] Trueman A. *The Coalfields of Britain*, (London 1954) p162
[9] Stephens, *Ibid*

The same source noted that the coal lay some 500-600ft above the Halifax Hard Bed. The Better Bed seldom lying below 130 yards and that, 'pits are sunk every quarter mile' alluding to the primitive mining methods being used. Here the Better Bed averaged 1ft 9ins in thickness and yielded 1,500 tons per acre. This coal was ideal for conversion to coke yielding between 65% and 75% from each ton of coal. The process took approximately 48 hours and its exceptional purity was said to contribute to the finished quality of the iron products. The same was not said of the iron ore found locally. The quality of this mineral was not considered to be very high. It was then the association of a good metallurgical coal and an average quality iron ore upon which was built the fortunes of the Bradford iron industry. Whilst it may have been lacking in quality this was compensated for by quantity; the iron ore was generally found in seam thicknesses of 22 ins or more and a single acre was capable of yielding up to 1,200 tons.

The Better Bed seams were worked mainly at depths of 240-250 ft in the east of the Dale and at 150-220 ft in the west. Overlaying this at depths of 30-100 ft was the Black Bed coal, considered too soft for metallurgical purposes but seen as a good thermal coal yielding between 4-5,000 cubic ft of gas per ton. It was also viewed as a reasonable engine and house coal[10]. This complicated picture illustrates how seam thicknesses varied throughout the District with faulting able to deprive some areas of coal altogether. Often the value lay in the close assemblage of minerals rather than the thickness of the coal seams being worked, which were much less than those to be found in the coalfields further to the east. For example iron ore could form the roof deposits of the Better Bed coal and the 'seat earths' might consist of valuable clays such as ganister utilised as a furnace lining material. In addition to these were the shallow outcrops of coal, a familiar feature throughout the District and providing an opportunity for

---

[10] Trueman, *op.cit.*p164

low cost working. Given the steep topography, these outcrops provided easy access and might form an even more basic level of mining activity.

Other coal types included the 'Thirty Six Yard' which alludes to the depth at which it was usually found. This was a coal mined in association with fireclay, a mineral produced as a result of the fossilisation of the roots of trees from the carboniferous era and used in the manufacture of heat resistant artefacts. Again this provides another economic linkage which might improve the prospects for coal mining. The 'Forty Eight Yard' coal which also takes its name from the depth at which it was found had been mined as a 'house coal' in Halifax, with the first references to the name occurring in the seventeenth century, a further indication of the pre industrial origins of coal mining in the District, where many small collieries were established to meet the domestic needs of isolated communities[11]. Even where the seams of Better Bed coal were only 1ft thick its value to the metallurgical industries was such that it was still economically viable to work. It was the presence of this coal type more than any other which was to promote the expansion of iron manufacture, the second most important industry in Bradford.

Fireclays and ganisters underlay and form the seat earths or floor of coal seams. These minerals also varied considerably in thickness with no relation to the overlaying thickness of the coal seam. They showed marked variation vertically in their chemical or physical properties, which meant that in thick fireclays it proved advantageous to work the beds in two or more lifts according to the change in composition, keeping the different quantities separate. The main components are hydrated forms of alumina and silica, fluxes such as lime iron oxide, magnesia and alkalis are also present but rarely make up more than 7% of the clay. The main value of this mineral was the property of refractoriness which meant it

---

[11] Transactions of the Midlands Institute of Mining Engineers (1950) edit 38

could withstand high temperatures. Working these seams then utilising the mineral for such purposes was often a matter of trial and error. Chemical analyses or other empirical tests may indicate the refractoriness of a fireclay without necessarily determining its suitability for a particular purpose. It was generally necessary to submit all untried clay to working conditions to see how it operated under the effects of pressure and gases. No single fireclay worked in the District possessed all the qualities required so normally each type of clay had its own range of end products. For example works were established to manufacture such items as firebricks, retorts, crucibles, sanitary ware, acid and fireproof goods. In addition it was also used for common household bricks, paring and ridge tiles. The mineral products of the District were being used both to build the homes as well as provide fuel for the expanding industrial population. Like coal other minerals were mined from shafts, drifts and quarries[12]. The small scale mining and quarrying of the pre industrial era, undertaken to meet basic domestic needs had provided the basis for experimentation and did some of the groundwork for subsequent economic exploitation. Much of this can be traced back to the quarrying done to provide the stonework for farmsteads, the evidence for which can still be seen in the weathered exposed rock on the fells.

Huddersfield was particularly well served by fireclays and as a consequence this proved to be an added incentive when sinking collieries. Fireclays with an alumina content which ranged from 2-30% were to be found among beds of the lower coal measures and the industry in Huddersfield spread over some 30 square miles, producing medium quality refractory goods. Apart from those for general building purposes, the fireclay produced the refractory bricks necessary for lining furnaces.

In Halifax there were two or more seams, Ganister or Hard Bed coal underlain by ganister and fireclay,

---

[12] Transactions of the Midlands Institute of Mining Engineers (1927)

separated by sandstone and siliceous shales. Most of the works making these goods were situated in the township of Elland, where the ganister reached the surface along the northern boundary with Huddersfield[13].

The driving force behind increased working of minerals and population expansion was the wool textiles industry. Initially it had been slow to move away from the use of water to drive machinery, in contrast to the more technologically advanced cotton industry in Lancashire. Ultimately the adoption of steam power was accelerated by the presence of the Better Bed seam and its reputation as a good engine coal. References to the availability of this coal and its value to the textiles industry first began to appear in the early nineteenth century. Its presence in a particular location could play a significant part in the decision where to locate as mills began to be established beyond the traditional valley bottom sites where, they had originally required close proximity to fast flowing streams.

There was no contiguous coalfield as such in the District. Instead the local mining industry developed in a piecemeal fashion, reflected in the fact that a variety of local names might be used for the same seam. Examples include the 'Shercliffe', 'Churwell Thin' and 'Beeston' all relating to the thin seams being worked in the localities which created these names. The 'Stone' coal was utilised for gas making in parts of Bradford and is presumably a reference to what was known elsewhere as Better Bed. The 'Blocking Bed' and 'Wheatley Lime' were favoured as both engine and household coals. Others included the 'New Hards' and 'Green Lane Old Hards'[14]. It is likely that some of these names were brought into being by the owner of a single colliery or perhaps a group of collieries as something of a marketing device. Each was to fill an economic niche in the District at a time when the only

---

[13] Searle A.R. *An Introduction to British Clays & Sands*, (London 1912) pp178-181
[14] Wray D.A. *The Geology of Huddersfield & Halifax*, HMSO (London 1930) p171

external source of supply was coal brought in by canal. At this stage during the early nineteenth century and before the arrival of railways, there was a need to rely on what was being mined locally to meet specific requirements. The 'Flockton Thick' for example was used for gas enrichment and illustrates how local industrialists were prepared to experiment with the different coal types available to determine their uses. In 1850 the owners of Lane colliery on the southern outskirts of Huddersfield attempted oil production using the Flockton coal, although the yield from three tons was said to have been only 20-24 gallons of light oil[15].

The vertical integration of industries using coal found in association with other minerals was often led by the producers of fireclay. In locations where the Hard Bed coal was worked this prompted the establishment of potteries linked to a particular colliery, producing non industrial artefacts for the local market such as flower pots, jars, pitchers and bowls. Even though the coal might be of inferior quality and found in a thin seam, it could still be utilised as fuel for the kilns. This shows how the development of both pottery and brick making in the District was closely connected to the expansion of the coal industry, with the same often primitive methods being used; in some cases either quarrying or 'day holes' as they were sometimes known. The latter being essentially short drives into a hill or valley side where artificial lighting was unnecessary.

The movement to green field sites by the wool textiles industry which began to gather pace after 1840 prompted the need for water supplies. Previously the velocity of the water supply had been the main consideration to provide energy but with the introduction of steam power the need for a reliable source continued albeit in a different form. The mining industry often supplied the expertise necessary

---

[15] Lees G.M. & Taitt A.H. The Geological Results of the Search for Oilfields in Britain, *Quarterly Journal. of the Geological Survey* (1946) pp255-317

for the drilling of wells. A by product of this was the discovery of small traces of natural gas. In an area where resources were not abundant this gave rise to more experimentation. For example, in 1845 a source of high pressure natural gas was successfully plugged and conducted to a mill boiler house where it supplied heat for several weeks [16].

It will be apparent that the majority of the various coal seams found in the District tended to be between 1-3ft in thickness. Their economic value was shaped by several factors. A seam of perhaps only 1ft in thickness would seem to lack viability but in the early part of the nineteenth century there was no general access to more abundant sources of supply which lay further to the west in the county. Because the railways were yet to establish themselves even the narrowest seams were worthy of exploitation, particularly if there were other minerals available to be worked. The major driving force was the wool textiles industry which as it became more dependant on the use of steam, meant the locational advantages of local coal became an important factor in the way mills were established throughout the District, particularly because the coal seams seldom occurred at depths of greater than 200ft and often much less, for example 80 ft being the average working depth in Huddersfield. In many colliery sinkings the technical expertise required in the early stages of development did not go much beyond what had been available in the pre industrial era. The steep sided Pennine valleys leant themselves to the easy working of coal and other minerals. Shallow pits of the kind that had been used for hundreds of years remained in use throughout much of the nineteenth century, where more sophisticated techniques would have been uneconomic.

Individually minerals found in the some parts of the District would have had only limited value, possibly none at all for industrial purposes. It was then economic

---

[16] Wray *op. cit.* p177

association of coal, fireclay, ganister and iron ore which combined to make mining viable on an industrial scale. In this respect understanding the economic geology means an appreciation of these linkages, chief among which being the low grade iron ore in Bradford Dale found in association with a low sulphur metallurgical coal making a vital combination. In other locations thin seams of coal could be worked in association with more valuable deposits of fireclay. The coal being used to fire kilns not only for the making of refractory goods, so important in many industrial processes but also allowing any surplus coal to be sold for domestic uses. These products sometimes linked with sandstone extraction were to be of value in building then sustaining the communities that were to live in the densely packed housing, vital for accommodating the growing populations of Huddersfield, Halifax and Bradford.

Another industry that came into being because of the availability of local coal supplies was that of dyestuffs. This was to transform the tar distilling trade. Various discoveries in the 1840s by Read Holliday who founded a firm of the same name in Huddersfield, led to a demand for such coal tar derivatives as rectified benzene for scent in soap manufacture and naptha as fuel for spirit lamps. In the early days of the industry crude coal tar was abundant. To obtain this it was usual to carbonise coal at relatively low temperatures, yielding both a rich gas and ample supplies of tar although this was to change as the demand for dyestuffs grew consuming more tar distillates[17]. The Holliday works located close to its customers also enjoyed proximity to several local collieries.

The ability to harness the local mineral wealth wherever it occurred quickly and usually without the need for significant investment, was the result of mining expertise accumulated over a long period during the pre

---

[17] Huber A. *The Chemical Industry During the Nineteenth Century* (London 1955) p55

industrial era. If industrial development had been required to await the arrival of the railway then it is likely growth would have much slower and expansion on to more distant green field sites something that would have had to wait until the supply problem was resolved. The economic geology provides the first consideration in the decision making process and might in itself be complex: the type of coal to be worked, the thickness of the seam, the association with another mineral that could be worked to greater benefit than the coal. These were the questions that provided the starting point when land was to be developed.

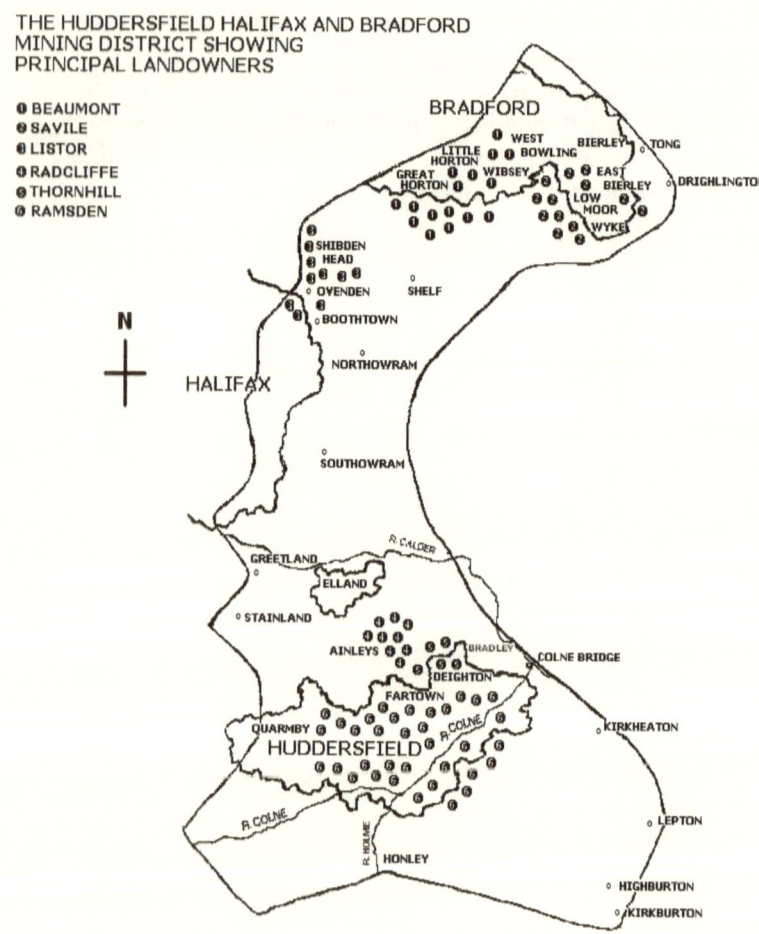

Chapter Two

The Landownership Pattern

The Introduction described how poor agricultural prospects encouraged an early recourse to other forms of resource exploitation. Throughout Great Britain the impact of the growth of industry impinged on the gentry, diversifying the nature of economic activities and income perhaps for the first time. Because of the long standing need to squeeze revenue from all available resources this was not the case in the District. Landowners were already well used to diversification and would stipulate a fixed annual rent 'certain rent' to prevent the property going unused and then a royalty on minerals extracted[1]. In this way responsibility for damage caused by mine workings was passed to the lessee. A complication was that the use of Ordnance Survey maps for denoting boundaries only came into practise during the twentieth century, a factor which was to complicate many disputes[2]. Further, due to the often small scale of mining activities and the limited financial resources that coal masters possessed, seeking compensation was often problematic. This could be a significant factor influencing the decision making process when coal reserves were to be exploited. For the major landowners in the District the choices could be complicated; providing an opportunity to utilise the coal reserves might be one of several options available to generate income. Often these options were to come into conflict and in doing so affected the viability of the coal industry and sometimes its ability to deal with competition

---

[1] Spring D. The English Landed Estate in the Age of Coal and Iron, *Journal of Economic History*. xi 1 (1951) p3
[2] Walmsley R.C. *Rural Estate Management*, (London 1948) p262

for the local market from producers outside the District. Another factor to consider was the risk of surface damage and the need to find space for waste disposal in an area where land use competition could be severe. Apart from having an adverse effect on land values this could mean less room for manoeuvre when reconciling an intrusive activity with other forms of land use; a further risk factor to take into consideration when considering what future use the land might be put to once mining activity had ceased. The effects of working an irregular mass of coal were not well understood in the early nineteenth century. The French mining engineer Fagol was later to observe that where coal was being worked at shallow depths such as those often encountered in the District, once an area of 350 m2. was attained then a hollow could be formed on the surface causing it to sink nearly 2ft in depth. In most cases the surface affected tended to be greater than the area of excavation underground[3]. Other effects could include vertical or horizontal movement, surface cracks or fissures and interference with the water supply. Such effects were best countered by the long wall system of working which left sufficient pillars of coal in place to help counter such a risk. However the long wall system was only attractive when large acreages of coal were to be worked. Small and primitive collieries often sank a series of pits and extraction could be unsystematic.

There were various ideas relating to the protection of buildings on the surface above mine workings. For example Merrivale's theory consisted of a formula where s equalled the length of a square pillar or the diameter of a circular pillar in yards and d equalled the depth of a shaft in fathoms thus: $s = 22\sqrt{d}/50$[4]. The extent to which such a scientific approach was applied was very limited. Estate surveyors grounded in agricultural practise would have

---

[3] Fagol M. Subsidence Caused in the Working of Mines, *Proceedings of the South West Institute of Engineers* (1896-97).
[4] Lane & Roberts, *The Principles of Subsidence and the Law of Support*, (London 1929)

had limited knowledge of how to manage the risk. The response varied. One of the largest landowners in the District was the Ramsden Estate. The property had originally been acquired in the sixteenth century by William Ramsden a yeoman farmer and by the nineteenth century the character of the estate began to change in response to industrialisation. However despite the estate manager lacking experience of anything other than agricultural practice, only rarely were experts in other fields consulted[5]. To a considerable extent the estate's response to industrialisation was largely passive. A flicker of interest had been shown by the construction of a cloth hall and subsequently the creation of the Ramsden Canal but nothing structural of note was put in place during the nineteenth century to assist economic development.

The properties of the Ramsden estate in Huddersfield and those in the immediate neighbourhood, do however provide an illustration of one response to the prospects offered for the mining of coal in an expanding industrial town. Coal reserves lay under much of the central area of Huddersfield and occasionally ancient mine workings were responsible for interruptions to building work. Quite possibly this alerted the estate's management to the risks involved when mining and building developments were in close proximity. This was especially true when numerous undercapitalised collieries were working on small leaseholds. It is of course reasonable to expect that the stewards of eighteenth and early nineteenth century estates had a greater knowledge of the management of agricultural land than urban development. Certainly the transformation of an estate whose income had traditionally been largely based on sheep and the cloth trade would have presented challenges.

From 1840 a policy was adopted which saw the gradual removal of mining from areas where building work was contemplated. This took the shape of a restrictive policy

---

[5] Spring *op. cit.*, p2

which refused to grant further mining leases within a one mile radius of the town centre[6]. Beyond this the estate was prepared to lease mineral rights but with restrictions. Leases would be granted only to coal masters with sufficient resources to create single site collieries rather than the more common series of crude pits. Standard terms were included in such leases; for instance there was a clause preventing sub letting. This may have been designed to prevent the small coal master from gaining access via another means. It was a common practice to boost production on a leasehold by the simple expedient of bringing in a sub contractor to sink more pits.

Surprisingly the requirement to make good any surface damage had not previously been a standard requirement, probably because many collieries had been established on low grade agricultural land where subsequent use beyond perhaps the grazing of livestock was not contemplated. The estate management was to learn the hard way what the effect could be on building projects when ancient and often unrecorded mine workings were uncovered on a site being used for building purposes[7]. This approach reduced the number of collieries operating on the estate, although those that did remain working between 1840 and 1865 had a longer life than was usual. It was more usual given the small size of leaseholds that a colliery might remain in operation for as little as five years. Seven of the ten collieries operating on the Ramsden estate during this period worked for between nine and twelve years and two continued until the turn of the century though they did not work continuously[8]. With activity concentrated on fewer sites and investment in equipment capable of raising larger quantities of coal, these collieries could produce between 8-15,000 tons per annum. Although there is no complete record of the income which the estate earned from mining, some indication is given by the fact that mining leases and

---

[6] Huddersfield Public Library, Ramsden Estate Papers R.A.4
[7] Ramsden Papers R.A.8.
[8] Ramsden Papers R.A.4.

royalties brought in £4,300 in 1857 and by 1858 this had risen to £9,700[9].

Whilst it can be seen that the policy allowed some collieries to be accommodated on the estate it did not necessarily work to the advantage of the wider mining industry in the town. Although mining activity remained at a high level between 1840 and the mid 1880s with as many as 38 collieries operating in some years, the full potential for coal mining in the town was never realised and the policy of the Ramsden estate played a significant part in curbing developments even on property it did not own. A degree of cooperation with neighbouring owners would have been necessary to fully exploit coal reserves and this the estate was not prepared to do. As the town expanded so the need for land not only to provide space for industry but also to house the growing workforce took precedence. The estate's correspondence reveals that requests for 'wayleave' (permission to move coal across a property) were regularly refused as were those for the construction of access roads and drainage tunnels. Similarly an ambitious plan to build a mineral railway linking collieries on the north eastern outskirts of Huddersfield to industry being developed to the south was rejected[10]. The failure to develop this scheme had long term effects. Though domestic customers were easily accessible, more valuable industrial customers notably the wool textiles industry were not so easy to reach. With the arrival of the railway local collieries found it increasingly difficult to deal with competition from outside the District. As early as 1855 there were 27 coal merchants operating from railway sidings in competition with local producers. This began the process whereby local collieries were edged towards the margins of the market.

Although only two miles distant from Huddersfield and under the same management the 1,100 acres of Ramsden property that lay around the township of Almondbury were

---

[9] Ramsden Papers R.A.6.
[10] Yorkshire Archaeological Society, Thornhill Estate Papers

treated differently by the estate. The properties in Almondbury were predominantly rural in character and there was little likelihood of substantial building developments taking place. Beneath the property was a 2ft thick seam of 'Soft Bed' coal which although easily worked lacked potential for long term development since there were no deeper lying reserves. However an industrial market for local coal was developing and by the late 1840s there were 14 woollen mills in the area, many powered by steam. Given these conditions the estate had no need to adopt the same policy as on the urban properties in Huddersfield and lots as small as two acres were made available for lease to coal masters who lacked the resources to do more than follow the wasteful and destructive mining practises common to the area. A typical colliery on the property consisted of a shallow pit or pits with a horse drawn gin or windlass to raise the coal. Once underground working had advanced to the point where it the need for roof supports arose, then the pit was abandoned and a new one sunk. Not only was this a wasteful method but underground hazards such as flooding often caused premature abandonment[11].

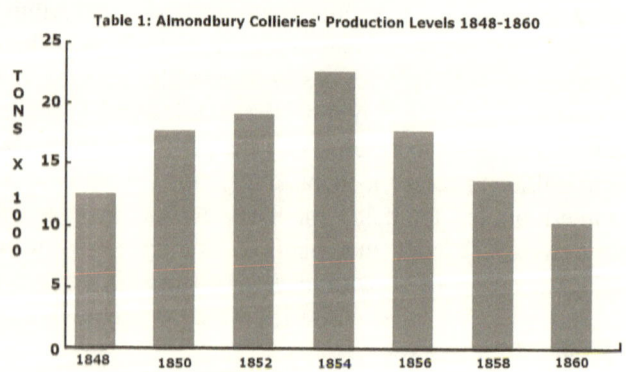

The year 1852 marked what was apparently the only occasion when a colliery engineer was asked to view

---

[11] Griffin A.R. *Mining in the East Midlands* London (1975) p113

mining operations. He reported that inefficient working methods were costing between £900 and £1,200 per annum in lost income. Despite this no change was made to existing arrangements. Probably the estate calculated that the nature of the coal reserves and the location made it unlikely that it would be possible to attract any but the small coal masters who already held Ramsden leases. It would seem that the estate was satisfied with the level of income generated by existing leases and royalties, though there must have been significant fluctuations as the selling price of coal varied from between 3s 6d and 5s per ton between 1835 and 1857. During that period an average of 16 collieries operated on the Almondbury property[12].

The financial records are incomplete but an account from the year 1852 indicates the sort of income individual collieries near Almondbury might generate. Close Hill Colliery which worked three acres paid a rent of £95 and royalties of £45. New Ground Colliery on a site of two acres paid £84 in rent and £40 in royalties[13]. These figures suggest that the contribution of individual collieries could be modest but collectively mining income in the year 1854 was £2,400[14]. This seems to have provided a significant boost to earnings on what was low grade agricultural land and where previously the only mineral extraction had been a few isolated coal pits serving the local domestic market, together with a scattering of stone quarries.

There were then significant differences between the management of the rural and urban properties on the Ramsden estate. On the former mining even on a small scale was permitted and even encouraged as a means of increasing revenues. On the latter it was severely restricted with steps taken to eradicate altogether the wasteful and erratic mining practices that were capable of damaging land seen to have potential for building projects. There was though one common feature: the estate had no direct

---

[12] Ramsden Papers R.A.5.
[13] Huddersfield Tolson Museum, Colliery Records, B.1.
[14] Ramsden Papers R.A.5

involvement in the exploitation of coal reserves. As a consequence no attempt was made to improve the infrastructure on the properties. Indeed the estate remained indifferent and the development of roads, drainage and other peripherals were left to the coal masters to deal with after negotiation with the estate management.

The Thornhill family of Fixby Hall owned a small estate on the north east outskirts of Huddersfield. The development of coal mining on this property illustrates the difficulties anticipating trends in an expanding industrial town. There were also properties at Stainland near Halifax but the mineral deposits here were considered to be too remote to be commercially exploited. As a result they never developed beyond an adjunct to agricultural activity and the focus remained on the Huddersfield portion of the estate. Up to the 1860s the quarrying of stone was the main extractive industry on the Huddersfield property. Then despite the estate having no experience of developing a colliery it went ahead with an ambitious plan to do so; the aim being to anticipate the growth of industrial and domestic markets in the neighbourhood. Direct development of this kind by a landowner was a rarity in the District. Thornhill Colliery was intended to be large by the standards of the area and it was anticipated that it would be capable of producing more than 10,000 tons per annum. The colliery came into operation during 1862. However the industrial developments anticipated did not occur. The area around the estate began to be favoured for the building of better quality housing for the expanding middle classes. Clearly this market would be inadequate for such a colliery. This prompted the attempt mentioned earlier to obtain permission from the Ramsden estate to construct a mineral railway allowing the colliery access to industrial customers in the Colne valley. The Ramsden estate had no interest in developing a railway which would bisect land that was already being utilised for house building[15]. As a consequence the colliery was unable to realise its full potential and eventually closed during the 1890s.

The Radcliffe baronets were landowners with widespread holdings in various parts of the country. The principal properties in Huddersfield were at Milnsbridge and Linthwaite to the west of the town. There were also other holdings in the Halifax area at Greetland and Elland. These small properties maintain the link in a continuous chain of mining activity on landed estates through the District. As with the Ramsden estate, correspondence reveals that these were agricultural properties whose stewards found it difficult to deal with mining questions and showed no interest in securing assistance, which is reflected by some of the long leases granted. In 1893 the estate leased 'clays, coals and stone' for 21 years to the William Robinson Brickworks. This was an unusually long lease for the District and effectively consigned the land to an extractive industry over a long period of time. What might have been a term of years better suited for quarrying was extended to include coal. It illustrates a somewhat clumsy approach to exploiting low grade and low cost minerals simultaneously. The brickworks would have been principally interested in the clays; coal would have been used to reduce fuel costs in the ovens and the stone sold as a building material. Similarly in 1853 the Jagger brothers well known coal masters in the District leased beds of coal for £30 per acre via an annual rent on a 21 year lease. Landowner's returns from 1875 report that Radcliffe's total landholdings amounted to 2,872 acres providing an income of £7,271 per annum, much of which came from mining.

Such long leases contrast with the approach taken on Radcliffe property near the village of Ainley between Halifax and Huddersfield . The entire property was agricultural and mostly given over to sheep and cattle

---

[15] Thornhill Papers

grazing. In the five years before the development of mining little income had been made from the property. Correspondence refers to the deteriorating condition of its buildings[16]. Evidently the estate was reluctant to spend money even on maintenance.

Beneath the property lay an 18in thick seam of coal. This was of only moderate quality. However the remote location coupled with poor communications made exploitation worthwhile. An initial attempt to begin mining in 1843 failed when local coal masters showed a reluctance to lease. Following this the estate was persuaded to embark upon the construction of three access roads to improve the transportation of coal[17]. The decision to make this investment suggests that the estate had finally begun to appreciate that some expenditure was necessary in order to improve the prospects of generating mining income from the property. Once these were complete however the maintenance became a charge on the estate and they were to suffer considerable damage over the years because of the heavy loads they were required to carry. The cost was a frequent cause of complaint in estate correspondence[18].

Once access problems had been resolved leases began to be taken up. Significantly the estate promised to bear a proportion of any drainage costs, a significant inducement on a site which had steep gradients. By 1844 four collieries were in operation working a total of twenty shallow pits. These earned the estate an income of £2,261 in that year. However expenditure on drainage tunnels known as soughs, together with other work cost £512. Small coal masters continued to be attracted throughout the 1840s. Typically they would employ between five and twenty miners. Depending on their financial resources the leases they took out covered plots of between five and seven acres. It was a precarious system and wasteful methods of

---

[16] Yorkshire Archaeological Society, Radcliffe Estate Papers 58:81
[17] Radcliffe Papers 58:28
[18] Radcliffe Papers *Ibid.*

working were common; principal among which was the pit, soon abandoned for another nearby. Coal masters rarely worked a lease for the full term. Correspondence suggests that they were often disillusioned by underground difficulties and quick to see the attractions of a new site[18].

Markets for the coal appear to have been restricted to within a three mile radius of the estate which accords with Nef's estimate of the market for eighteenth century collieries[19]. This is not surprising given the remoteness of the site. However by the end of the 1840s three woollen mills had been established in the vicinity taking up to two thirds of the estate's output which had averaged 38,000 tons per annum in the last half of the decade. The remainder would appear to have been sold as house coal and to small workshops operating on the outskirts of Huddersfield. With these developments on the estate came benefits for the adjacent properties. The new roads allowed other coal masters to negotiate rights of wayleave and open sites where access had previously been difficult. In 1842 production on neighbouring properties was 11,000 tons. By 1849 this had risen to 32,000tons. Taking advantage of improvements in infrastructure on the Radcliffe estate meant that other local landowners were able to enter the trade. However as production on the estate declined so this was mirrored on neighbouring properties, despite continuing investment. Up to this year an average of £200 per annum was spent on improving or maintaining infrastructure directly beneficial to lessees in an attempt to sustain mining activity[20]. This may have been prompted by leases being made available on neighbouring properties in more attractive locations. A typical lease on the estate would be £30 per acre annually

---

[18] Springett R.F. *The Mechanics of Urban land Use Development 1770-1911* unpublished. PhD. Thesis (University of Leeds 1979)
[19] Nef J.U. *The Rise of the British Coal Industry* (London 1932) pp102-109
[20] Hartley W.P. 'Five Landed Estates in Yorkshire and the Development of Coal Mining', *The Local Historian* vol. 23 (1993).

for each bed of coal and not less than £20 per annum 'whether coal in quantity equal in value to that sum after the rate aforesaid shall be gotten'.

Gradually expenditure drew closer to a declining income. To some extent the true financial situation had been disguised by wayleave payments and the occasional compensation for encroachments from neighbouring properties, since no attempt was made to differentiate between the different sources of income in the estate's accounts. Only in 1860 did the estate appreciate the full extent of the decline in mining revenues. During that year a fall in local demand reduced wayleave payments and exposed the true position[21]. After 1860 with the decline continuing it became impossible to sustain the needs of the few industrial customers. Whilst the larger leaseholds had not been entirely exhausted, the effects of wasteful mining practices made it uneconomic to continue. Even at significantly reduced rents no-one could be persuaded to lease what remained. With the estate now showing less interest in the maintenance of the infrastructure, neighbouring properties also found it harder to move coal even to local customers. Income from mining continued to fall (it was down to £220 by 1862) with production continuing only spasmodically. By 1865 estate accounts made no mention of income from mining sources. Neighbouring collieries suffered a similar fate: by 1868 there were only two operating within a two mile radius of the estate. The dual approach to the way in which mineral rights were disposed of by individuals lacking the necessary technical knowledge poses the question was this simply a 'best guess' method shaped by the market conditions; the likelihood (if any) that the land would be put to a more lucrative use or even that competition might encroach reducing the value of the coal to the extent that mining might become uneconomic? The latter was an important consideration since by the 1850s it was already

---

[21] Radcliffe Papers 58:26

apparent that the arrival of a railway branch line in an isolated location could have the effect of rapidly bringing about the demise of local mining[22]. These and other factors will be explored in due course in order to establish what strategies were at work throughout the District. At this stage it will suffice to note that unlike the Almondbury property of the Ramsden estate which though rural was far from isolated, some Radcliffe holdings suffered from remoteness and steep gradients. This was land that might best be described as 'intake' from the surrounding moors, absorbed during the eighteenth century and in contrast to the settled holdings around Almondbury, where roads and drainage had been created incrementally over many years and could therefore be adapted to assist the development of mining. To tempt coal masters onto such an unpromising site needed investment and it is unlikely that the estate would have become interested had it not been considered necessary to stimulate activity. Shallow coal measures meant that the working life of any colliery on the estate was likely to be short. The small coal master and his methods accelerated this process. Whilst the neighbouring properties shared the same problems of isolated sites and steep topography none was prepared to take any direct action to encourage mining activity, probably because the cost would have been too great and the returns uncertain. In taking the lead the Radcliffe estate stimulated activity to some extent on the neighbouring properties. By subsequently neglecting the infrastructure it had created, the estate helped bring about the demise of mining elsewhere in the locality.

The Lumley Savile family owned property around the township of Shelf between Bradford and Halifax, amounting to a total of some 800 acres. Before mining was established, agriculture had formed the main source of income. A scattering of small farms kept livestock and grew crops. In 1838 the total income of this estate

---

[22] Hartley *op. cit.*

amounted to £650. Legal constraints prevented the exploitation and development of coal deposits until 1840 when a Private Act of Parliament allowed leases to be granted. The Bowling Iron Company of Bradford took up the first of these in 1842, beginning the partnership between landowner and manufacturer. As ironstone occurred in association with the coal these deposits were particularly attractive to the company. The Bowling Company established collieries of its own where the size of the leasehold was likely to repay investment in a sophisticated mining operation capable of producing over 10,000 tons per annum. On smaller leaseholds they preferred to engage sub contractors or Pit Takers as they were known locally, to establish and then work rudimentary collieries. Transport to the works was via a network of mineral railways. Not surprisingly this activity began to change the face of the estate. The first lease of 1842 earned the estate £276 in that year. This was a modest beginning. By 1847 five more leases were earning £4,342. Income continued to grow and by 1849 it had risen to £9,000[23]. Mining had by then become the dominant activity on the property and was overwhelming other forms of land use. By 1850 income from agricultural activity had fallen to £210.

Exploitation of the Lumley Savile property was unsystematic; the first three collieries on the estate were a hastily sunk series of pits working shallow deposits. It is likely that the company exploited these seams for ancillary purposes since the coal was of an inferior quality and unsuitable for use in a metallurgical industry. The deeper deposits were far better for this purpose. Deep mining techniques began to be introduced on the estate after 1845 working alongside the primitive pits[24]. The characteristic headframe was raised over a shaft and steam power used for winding purposes. With some seams lying at depths of up to 300ft this method was clearly necessary. At the end

---

[23] Dewsbury Public Library, Savile Estate Papers, B101
[24] Savile Papers *Ibid.*

of the decade nine collieries were in operation and this number remained constant up to the 1870s, although old sites were abandoned and new locations developed. After 1850 production was maintained at over 60,000 tons per annum until 1875 when there was a significant decline down to 40,000 tons. Thereafter production continued to fall and by 1885 it was below 10,000 tons.

The Savile estate enjoyed the advantage of proximity to a large industrial consumer who was prepared to pay generously for the locational benefits of extracting coal and ironstone. The estate was in the main content to simply lease but there is evidence to suggest that it was prepared to act on behalf of the company. The property did not consist of a single block of land. Smaller holdings belonging to other owners lay within the estate boundaries. As long as agriculture was the common enterprise no significant problems seem to have arisen. When however, neighbouring landowners attempted to exploit their coal reserves the estate began to assert its dominant position, usually by refusing to grant wayleave. During the 1850s several landowners in this situation wrote to the estate requesting to use Savile property as a means of facilitating mining developments. Each time these requests were unsuccessful[25]. In effect the estate was refusing to assist in the development of properties where the owner wished to find a different market for the coal. Subsequently these landowners made arrangements with the Bowling Company. This provides an illustration of how close the relationship was between the company and the estate. In due course the Bowling Ironworks became the major promoter of mining activity within the area.

The close relationship between the Savile estate and the Bowling Company was mirrored by that of the Low Moor Ironworks and the Beaumont family of Bretton Hall near Wakefield. However Low Moor was unusual in that the company was also a landowner in its own right. During the

---

[25] Savile Papers

eighteenth century the value of minerals in the area was first appreciated by Edward Leedes a local landowner. Whilst he was undoubtedly the visionary it was the company who realised who realised the potential of his land. Leedes proved to be a poor businessman and the company was to buy up the Royds Halls estate after Leedes went bankrupt[26]. This property was to provide high returns from the iron ore mined there in association with the Better Bed coal. By 1801 the Low Moor company was the largest single landowner in the township of North Bierley paying nearly 70% of total land tax assessment. Interestingly and in common with the other landowners, the company was to retain many of the practices associated with an agricultural estate[27].

As on the Savile estate the process of exploitation often began with small leases of five acres or less on the 550 acre Beaumont property. This suggests that the company was at first cautious in its approach, wishing to ensure that yields were sufficient. Ultimately no part of the property was spared from exploitation, although during the period from 1840-45 there were others working leases there too. In 1846 a fall in the price of coal caused some local coal masters to abandon their leases. These were then taken up by the Low Moor Company which rapidly became the dominant leaseholder on the estate. Reworking these abandoned collieries created a great deal of surface damage, something the estate tried belatedly to limit. The company agree to reduce the number of shallow pit workings and concentrate instead on deeper mines[28]. This resolved one matter but another problem soon arose. On some parts of the estate the company held leases to mine coal on others ironstone and elsewhere leases for both. This confusing system was allowed to evolve unchecked for many years until in 1874 a survey was made to clarify the situation. This revealed that coal had been mined

---

[26] Bretton Hall, Beaumont Estate Papers, B11/45a
[27] *The Development Bradford Iron Working 1500-1830* (WYCC 1974)
[28] Beaumont Papers, B11/53a

illegally in a number of locations. Small coal masters acting as sub contractors were held to blame and the company paid £9,300 compensation to the estate[29].

The Beaumont estate took an entirely passive approach to the exploitation of coal on its land, preferring to do no more than grant leases. Income could be considerable. For an acre of coal in Bradford in the 1840s £30 per annum was the usual rate. Additionally the Low Moor Company agreed to a guaranteed payment of £600 per annum on the larger leaseholds irrespective of whether the coal was worked or not. A large industrial consumer with a voracious appetite for coal was prepared to pay such sums to secure local supplies. It should be noted that such payments might also cover the right to mine iron ore. This close connection with the estate was to have long term benefits. In the 1880s as the coal reserves on the Bradford leaseholds began to be worked out so the company was able to secure supplies from other Beaumont properties near Barnsley[30].

Whilst mining was being undertaken in other locations around Bradford, it was the demands of the iron companies on the southern outskirts of the city where the greatest levels of activity took place throughout the nineteenth century. This was reflected not only in production but also colliery numbers. Because the bulk of the coal was being sent to two industrial customers and even that which was not used for iron making was sold to employees or used in associated activities, it is not possible to determine accurately just how much coal was being produced annually for the purposes of iron manufacture.

To the north east of Halifax in Shibden Dale lay the 450 acre Lister estate. Poor agricultural prospects and little opportunity for building developments made mining an attractive proposition for the estate. During the early 1840s the collieries consisted of primitive pits which were no

---

[29] Beaumont Papers, B11/53a

[30] Beaumont Papers, B11/45a

different to those on neighbouring properties. Their market was local and mostly domestic. Colliery records however reveal a gradual effort to improve working methods[31]. Purchases were also made from neighbouring collieries to augment coal mined on the estate. Gradually the primitive pits were abandoned and the various sites coalesced to just three, each with a shaft sunk to reach the 'Better Bed' coal. Improved technology also appeared, notably steam powered winding machinery and with this an increase in the workforce from 51 in 1847 to 127 in 1850[32]. By 1860 the collieries were part of an integrated enterprise which included brickworks and a sulphuric acid plant[33]. One of the reasons why the Shibden Dale estate was able to establish itself as the town's only mining operation producing coal for industrial customers on a relatively large scale was poor transport links. Since railway penetration of the local market proceeded more slowly than in Huddersfield and Bradford, the estate was able to take the risks involved in making investments in modern mining methods.

There were then seven significant landowners in the District; aristocratic, gentry and corporate, though the latter held land in the same manner as an estate owner. This chapter has begun to identify more of the complex factors involved in the decision making process when mining was contemplated; in some cases made more difficult because management lacked the necessary expertise and often seemed unwilling to acquire this. Among these might be time constraints; the need to win coal quickly before an encroaching railway network could render uneconomic an isolated coalfield with thin seams. A shallow coalfield could be exploited rapidly using primitive mining techniques but if these were adopted then

---

[31] Halifax Public Library, Lister Estate Papers, CMA1
[32] Springett RG *Yorkshire Mill Town: a study of spatial patterns and processes of urban & industrial growth in Halifax 1801-1901* unpublished PhD. Thesis (Univ. of Leeds 1974) p161
[33] Lister Papers, CMA4

the landowner would have to accept that such methods were wasteful and much of the coal would be left unworked. Space could also be a consideration, particularly if the mineral rights lay in or close to an expanding urban area. Even a small primitive colliery would create the need for land to be set aside for waste disposal. Some landowners remained detached from the process, simply leasing mineral rights. Whilst this could bring significant returns via rents and royalties, a lessee whose primary business was not mining might show insufficient interest in how and where the coal was being worked. For others there was the challenge of direct involvement; the potential for gaining access to industrial markets that would make investment worthwhile, or alternatively the rewards that could come from vertical integration.

Against this background of complex decision making lay the wider industrial economy. The Heavy Woollen District needed coal to fuel industrial expansion. Between the late 1830s and the 1850s before the railway network had coalesced, the local mining industry was required to bridge the supply gap and had to do so within the constraints placed upon it by those who owned the land.

So far two broad elements in the decision making process have been revealed. The first being the economic assemblage of minerals. A thin coal seam might in itself be an unattractive proposition for mining but such factors as coal type, location and association with other minerals notably fireclay, might change this view. Within the geological question there were subsidiary factors, for example a thin seam of coal might be of use if the fireclay was of low quality and therefore more suitable to be used in situ for brick making, where the coal could be used to heat the kilns. There was also the prospect of extracting coal tar derivatives and pyrites, which again could increase the value of a thin seam. The most notable example which was to dominate mining and landowner decision making was that between the Black Bed ironstone and the Better

Bed Coal in the vicinity of the Bradford ironworks.

The second element was the destructive nature of mining. In areas where the land was of only marginal value and the potential for other types of use limited, then this was not important. However where there were prospects for other uses notably building, then mining presented a dilemma; the value of the coal and other minerals to be extracted set against the risk of surface damage and space required to dispose of waste. In addition there was the limited expertise of owners' agents who were more used to managing agricultural estates. Facing this were the demands from industry, notably wool textiles that required both land for mills and coal to fuel expansion. During the early stages of expansion the ability to source coal locally was vital.

Chapter Three

The Role of the Wool Textiles Industry

During the early nineteenth century the change from water to steam power gathered pace in the wool textiles industry. In turn this increased the demand for coal making it the largest industrial consumer in the District. During the 1840s and 1850s the lack of an effective transport network restricted the market area of collieries and in turn this was to influence the location of woollen mills. More will be said in due course about the evolution of the railway network and its effect on the coal mining industry. At this stage it is sufficient to note that it was not until the 1860s that a network of branch lines had evolved sufficiently to reach all of the once remote locations in the Pennine valleys. It was not surprising therefore that many mill owners were closely involved in the development of the coal mining industry.

One of the most famous names in wool textiles was John Foster who built the Black Dyke Mill at Queensbury near Bradford. Foster provides an excellent example of how closely intermingled was mining, the weaving of textiles and agriculture. It was this association which was to be of assistance in the development of both industries. A further advantage was that Foster owned land beneath which lay coal and local mining expertise was available. Land in the Queensbury district was of limited agricultural use and consequently the sinking of shallow pits had long been an additional or alternative form of land use[1].

From the 1830s the names of small coal masters involved in supplying the Bradford textiles industry began to appear; Fletcher Easby and Company, Hodgson Newall,

---

[1] Brotherton Library, Leeds University, John Foster, Business Records

Farrar, Ingham and Halliday were coal masters involved in what was a haphazard and competitive trade. There is much evidence from the business records of the mills to show random purchases of coal from a variety of sources. During the 1850s the average cost of engine coal was four shillings per ton and invoices to mills could range between £10 and £50. Additionally merchants were contracted by the mill owners to enter the market on their behalf[2]. It is likely then given the small size of their collieries, that many coal masters were unable to meet demand from their own sources and in order to maintain business acted as merchants for the mills, seeking supplies from further afield.

Foster began his involvement in the coal industry with a series of small pits with names such as Croft Ing Cottages, indicative of the close connection with agriculture. These were typified by the usual wasteful methods of working with much coal being left in place to support buildings. Between 1837 and 1863 two acres per annum were being mined on a 36 acre site. Competition from the iron companies and the increased demand from his mill caused Foster to look further afield, leasing mineral rights in the township of Northowram and employing pit takers to do the work. The connections with agriculture continued, for example extending the practice of gifts to workers at New Year to include the miners he employed.

Foster's Black Dyke Mill became a noted producer of mohair and worsted goods with the premises being situated on the site of the family farmstead. It was to become the largest employer in Queensbury with 700 weavers. Power looms were introduced as early as 1836 and eventually engines totalling 1,600 horse power were at work in the Victoria Mill, which by 1865 also had its own gasworks and was consuming 15,000 tons of coal

---

[2] West Yorkshire Wool Textiles Industry: A Catalogue of Business Records, Composite Ledger, 1855-62

annually[3].

Although Foster could be seen as something of a pioneer he did start with the advantages of land, the ownership of coal reserves and doubtless some knowledge of mining because of its close association with agriculture. In general the wool textiles industry was slow to take up new innovations unlike the Lancashire cotton spinners[4]. However influences from across the Pennines meant that by 1838 the take up of steam power was rising rapidly and by 1850 mill building accelerated in sites away from riverside locations. From this point onwards the wool textiles industry catering as it did mostly for the home market continued to expand.

In Huddersfield the move to green field sites utilising local coal supplies to support the installation of steam power, is illustrated by George Crossland who built a mill on land he had owned at Crosland Hill. By 1847 he is recorded as owning most of the cottages in the district where his workforce was housed. Here we see a once isolated agricultural community with activity previously based on agriculture, domestic weaving and mining, becoming an urban industrial settlement[5].

The presence of local coal supplies whose exploitation was often undertaken by mill owners, both enabled the rapid take up of steam power and allowed the industry to expand without waiting for the railway network to catch up. In turn this gave the industry the confidence to expand further upon the outbreak of the Crimean War. Whilst peace brought an excess in capacity, the ability to take up new developments in technology, underpinned by steam power meant that real costs were reduced and the industry was soon back to full employment by the middle of the

---

[3] Smith A. *An Historical Introduction to the Economic Geography of Great Britain* (London 1949) p213.
[4] Crump W.B. The Leeds Woollen Industry, 1780-1820, *Transactions of the Thoresby Society* (1929) p25
[5] Dyas H.J. *The Study of Urban History* (London 1972) pp221-226

decade[6]. Following Rostow the industry utilising water power took off in the years 1770-1800, the drive to maturity followed after 1830 and was to last for half a century. In 1861 the number of power looms stood at 11,405. Six years later this had risen to 20,713 and by 1874 had reached 30,917[7]. This rapid growth much of it taking place as it did in the years before railway expansion was complete, illustrates the crucial role played by the local mining industry.

The take up of steam might have been much slower given the reliability of water as a source of energy in the Pennine valleys. In the case of water power it can be said that every site was unique and that there were no standard sets of costs for operation and maintenance, setting the consistency of water power against high fuel costs and technologically unreliable steam engines, so at first glance it might seem wise to retain water power in the Pennines as one writer suggested but then made the further point that the cost of fuel was cheap[8]. Cost however needs to be understood within the context of control and accessibility and as has been noted many mill owners were also producers of the coal burnt in the engine houses. These advantages coupled with primitive mining techniques used in working limited acreages, were the drivers behind the expansion of the industry.

The consumption of coal by textiles and allied trades rose rapidly. By 1871 a weaver was estimated to burn 12 tons of coal each day for each boiler, dyers and bleachers 19 tons. In addition specialist machine tool manufacturers such as Butler and Skirl both Halifax firms, were operating in the District[9]. Most significant of all in terms of ancillary

---

[6] Church R.A. *The Great Victorian Boom 1850-1873* (London 1975) p41

[7] Glover F.J. 'The Rise of the Heavy Woollen Trade of the West Riding of Yorkshire in the Nineteenth Century', *Business History* vols. 4-6, 1961-64 p287

[8] Chapman S.D. The Cost of Power in the Industrial Revolution, *Midland History* vol. 1 (1973)

[9] Saul R.B. *The Development of the Mechanical Engineering Industries*

trades was perhaps the firm of Read Holliday established in Huddersfield in1850. The company created aniline dyes and in doing so began the organic chemical industry.

Foster opened his first colliery Spring Head in 1832, initially to serve the local domestic market. Originally the colliery had been a series of scattered pits employing a dozen or so miners. In 1836 when the colliery made a profit of £172, work began on modernising the site. A brick lined shaft was sunk over which an iron head frame was erected. This provides an early indication that Foster was making a serious entry into the coal trade and investing for the long term. Subsequently steam powered winding machinery was added and eventually the other pits were abandoned and the colliery began to operate from a single site. For Foster there must have been a considerable element of risk involved in this. Utilising mineral resources via the sinking of shallow pits was a low skill operation. However developing a coal mine in the fullest sense of the term, bringing with it all of the additional on costs associated with such a venture, was an ambitious undertaking for a man associated with farming and weaving. The colliery was now in a position to achieve and sustain consistently high levels of production. By 1840 these were at a level whereby output could be sent to the mill. In 1846 annual production was 4,020 tons. Two years later this had risen to 6,975 tons and in 1846 had reached 8,009 tons[10]. Even so Spring Head colliery was still not producing enough to meet the demands of the mill engine house and in 1842 Foster began to expand his colliery interests.

During the 1840s mineral leases to the south of Bradford became increasingly hard to obtain, the market being distorted by the demands of the Low Moor and Bowling iron companies. There was fierce competition between the two for mineral leases and often their

---

in Britain 1860-1914, Economic History Review series 2 xxx (1966) p126.
[10] John Foster, Business Records

financial strength enabled them to exclude other parties. Such was the demand for coal and iron ore that leaseholds of as little as two acres were of interest. The effect of this was that mill owners and other industrialists seeking coal supplies often had to look further afield. John Foster for example, finding that his own mineral rights were inadequate turned to Halifax, entering into a partnership with a local coal master named Lacey. This proved to be a beneficial arrangement. Three collieries were subsequently developed; Nursery (1842), New Keelham (1844) and Clews Moor (1847). The first of these though leased as a going concern had in fact been standing idle for some months. Correspondence between Foster and his partner refers to the amount of work necessary before mining could begin again. The impression is that Foster had misgivings about the enterprise but evidently could not spare sufficient time from his textiles interests to give the matter sufficient attention. This was a problem that occurred frequently when businessmen who lacked sufficient technical knowledge went into the coal trade. The need to secure an exclusive local supply of coal often dominated their thinking and could have unforeseen results. This was particularly the case when sub contractors or pit takers were employed. In their wake went complaints about surface damage and illegal encroachments onto neighbouring properties.

It transpired that parts of the colliery were flooded and it was necessary to drain these before further work could be done. After this was achieved the sum of £55 had to be spent on timber and other materials to make the tunnels and roadways safe. By the end of 1842 production had resumed. In 1843 a profit of £109 was made. This rose to £271 in 1846. Thereafter production fell and by 1848 profits had declined to £64[11].

Recurring difficulties with flooding helps to explain the decline. This was a hazard found in many collieries

---

[11] John Foster, Business Records

working such a shallow coalfield. Rather than use the expensive option of steam pumps to drain the mine an alternative and it was hoped permanent solution to the problem, was to dig a drainage tunnel or sough. Given the hilly topography of the District this was a promising though time consuming solution and could entertained only when the amount of coal to be worked was worth the effort and would therefore repay the expense. The fact that nearly 20 acres of coal remained unworked was sufficient to cause this method to be used. Foster had already noted how much of the mine's potential profits were already being used as fuel to operate pumps. Progress was however slower than expected and after six months the work was abandoned.

Whilst involvement in this and other ventures would have helped reduce fuel costs at the Black Dyke Mill no effort seems to have been made to coordinate supplies or even anticipate demand. In theory at least demand could have been linked to output but there seems to have been no effort made to achieve this. The engine house simply utilised what had been produced.

The partnership appears to have been nominal at best with Lacey concentrating on colliery related matters and Foster being satisfied providing some coal was reaching the mill engine house. There are many examples of demands being made by the mill, often at short notice for additional supplies. The impression is that the whole process was poorly coordinated[12].

In the Holme valley to the west of Huddersfield, a mining industry had existed for at least two hundred years. As older mills occupying valley bottom sites began to install steam power and new mills were opened, the value of local coal supplies grew in significance. A rail link was not established until the middle of the 1850s and as a consequence this isolated location had been heavily dependent on local supplies of coal during the first half of

---

[12] John Foster, Business Records

the decade, as mill owners began to install steam power. This location is of particular interest since it represents in a microcosm the challenge faced by mill owners throughout the District. In Holmfirth there was no canal link, the railway was yet to arrive and it had become apparent that steam power was the future. Local mill owners sought to acquire an exclusive local supply in preference to buying all their needs on the open and at the time rather limited market. Considering that the local mining industry had grown up to meet domestic needs, it was inevitable that such involvement would be highly speculative.

In 1852 Thomas Dyson owner of the Royd's Mill at Holmfirth leased a ten acre site on the recommendation of a local miner. This provides an illustration both of the lack of expertise available and in the scramble to acquire mineral rights, mill owners would seek the opinion of anyone who appeared remotely qualified to assess the prospects of a leasehold. Various difficulties followed, not least of which was the removal of coal from a site that lay on a steep slope. This had not been anticipated and Dyson found it necessary to construct an access road with a reasonable gradient for horse drawn traffic. Before any coal had been worked Dyson incurred expenses of £182. This included the cost of building the road and driving a drainage tunnel. It was a considerable outlay in order to work such a modest acreage and was well in excess of the start up costs to establish a mine on similar leasehold in nearby Huddersfield which often averaged less than £100.

THE COLLIERY DISTRIBUTION PATTERN
IN THE HOLME VALLEY 1845

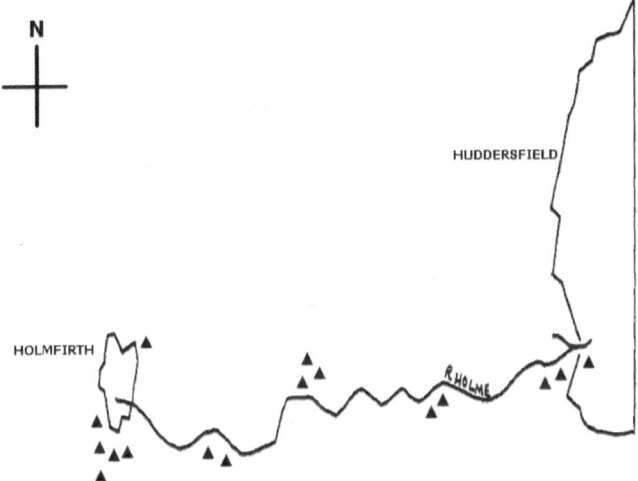

By 1853 450 tons of coal had been mined but then an area of what was described in the colliery day book as 'barren ground' was encountered. This was a problem often encountered where geological conditions had brought about a break in the continuity of the coal seam. The colliery was then abandoned. Evidently Dyson was not prepared to accept the risk or expense associated with driving through unproductive ground in the hope of encountering a workable coal seam once more. The coal mined had a market value of £90 set against total expenses of £383[13].

By 1840 there were 26 small collieries operating in the Holme Valley. Most were simply shallow pits or day holes

---

[13] Huddersfield Tolson Museum, Colliery Records, B.1.

driven into the valley sides. Production varied widely. Four collieries were said to be producing in excess of 8,000 tons per annum. Others produced as little as 1,500 tons per year. During the period when the number of steam engines in local mills was increasing and before the arrival of the railway, these collieries were vital in maintaining an affordable supply of coal. Without them it is difficult to imagine how mills working in a relatively isolated location could have made the transition from water to steam power, much less expanded in numbers, without a local coal supply. Collieries with such modest output in some cases probably not much more than was being produced previously for the domestic market have been given scant attention. Whilst this question will be addressed in greater detail in due course, it is worth noting that the *collective* output of these collieries was what mattered. It was not the only reason why the small primitive colliery was of such value to the wool textiles industry. However in the early stages of expansion it did play an important role. Figure 1 shows that mining activity in the valley mirrors the main sites of industrial activity, chiefly wool textiles with all of the collieries lying within a half mile of their markets.

Brooke's Mill at Armitage Bridge between Huddersfield and Holmfirth had two small collieries at work during the years 1837-41. The collieries were based on land owned by the mill. This was a fortuitous occurrence. Many of the older established mills owned or leased land which was once used for tenterhooks; the drying of finished cloth. The mineral resources lying beneath such property could be an additional bonus and another reason why mill owners moved into mining. Here the mill appeared to have hired a pit taker or at least someone claiming to have the necessary experience. Certainly he was well recommended and had worked in the trade locally for many years. During this period the collieries produced between 1,800 and 3,050 tons per annum[14]. The coal was won from a seam little more than 1ft in thickness and would have been of little economic value in a less isolated location. Although engine house records at the mill do not reveal whether these collieries were responsible for meeting all requirements, it is unlikely that this was the case. Even where output theoretically matched demand, such primitive collieries were prone to interruptions in working and it might take days to get back into production. The use of a crude pit for example, would mean that eventually a stage would be reached where abandonment became necessary for safety's sake. Coal masters were not enthusiastic about augmenting the workforce to sink a new pit whilst the old one was still in production. Such collieries often operated on the margins. The latter is illustrated by the fact that when a railway branch line passed through Armitage Bridge in the 1850s the collieries quickly closed, being unable to compete with low cost supplies delivered from outside the district.

In 1844 the owner of Newtown Mill which lay on the Bay Hall estate, part of the larger Ramsden estate, leased mineral rights at £40 per acre. When he proposed to open a colliery the services of the Ramsden estate surveyor were loaned. The surviving weekly reports detail the progress made in sinking a shaft and developing the colliery[15]. It is significant that the estate surveyor should have become involved, since this reveals another example of how collieries were established under the guidance of people who lacked the necessary specialist knowledge. It is to be presumed that the surveyor would have had his employer's interests as the priority rather than those of the mill owner.

By 1845 the colliery had come into production. Despite working a thin seam under the usual primitive conditions, the miners managed to earn an average daily rate of 4s

---

[14] Hartley W.P. *Involvement in Coal Mining by the Wool Textiles Industry: Some West Yorkshire Examples.* Yorkshire Archaeological Journal vol. 65 (1993) p169.
[15] Huddersfield Tolson Museum, Newtown Mill, Business Records.

during that year. This was 6d above the national average for the time. The day book provides an insight into the running costs for a colliery of this type. During 1845, timber props were purchased at a cost of £3 6s rails for underground haulage at £4 15s. Work done by various tradesmen amounted to £15 3s 2d. The wage bill came to £180 and total costs for the year were £231 6s. Some items such as corves or tubs for transporting the coal underground were manufactured in the mill workshops which probably helped to reduce costs[16].

It is difficult to be precise about the value of the colliery to the mill. In 1846 production amounted to 940 tons which on the open market at the time would have cost £235. Whilst it would seem that there was little financial advantage in the mill mining its own coal, the benefits would come from a reduced requirement to buy fuel on the open market. The railway network had not yet made an impact and it is likely therefore that Newtown Mill would have faced periodic difficulties in acquiring supplies in the right quantity at the precise time they were needed. Access to an exclusive supply reduced the risk of a shortfall.

An attempt was made in 1847 to expand the colliery by leasing a further 15 acres. However the Ramsden estate refused to allow any extension of mining. This is an example of the Ramsden estate policy mentioned in chapter two being brought to bear. The mill lay on the fringes of an expanding urban area and complaints had already been made about subsidence. The surrounding property had potential for building developments and the estate was not prepared to risk surface damage and subsidence interfering with the revenue that would be gained if this work took place. Despite this the colliery continued in production until 1851.

Another Huddersfield firm Middlemost Brothers purchased a colliery close to their Clough House mill in 1842. During that year the mill used an average of 344

---

[16] Tolson Museum, Newtown Mill, Business Records.

tons per month. The colliery, New Haigh, produced an average of 103 tons per month[17]. This provides an example of a textiles company buying a colliery as a going concern and some effort seems to have been made to secure professional advice. This occurred in 1843 when attempts were made to increase production. A mining engineer inspected the colliery and recommended various improvements; notably the installation of steam powered winding machinery for raising the coal and the sinking of a second shaft lined with bricks. References in his report about the use of wicker corves for moving the coal underground and inadequate timber for roof support, gives an indication of the primitive methods in use[18].

Middlemost Brothers chose not to act on the recommendations. They did however hire six more miners in an attempt to increase production. Correspondence suggests that the firm was not confident of being able to acquire more mineral rights to justify investment in new works and equipment. New Haigh colliery supplied a total of 7,800 tons of coal to the mill. Annual costs including wages never exceeded £350. During this period the price of Soft Bed coal in Huddersfield averaged 5s 6d per ton. This meant that the firm received coal to the value of £2,145.

In Halifax mill owners had been slow to adopt steam technology but by the 1830s there had been a huge influx of power looms into the industry greatly increasing the demand for coal. Mills in the Hebble valley were faced with similar problems to those in the Holme when securing fuel supplies; isolated locations with only a few small collieries in the immediate vicinity.

A limited solution came when the Thornhill family of Fixby Hall near Huddersfield decided to lease mineral rights close to the village of Ovenden. This decision can be linked to developments on their Huddersfield property mentioned in chapter two. At that point the estate saw

---

[17] Leeds City Archives, Middlemost Brothers, Business Records
[18] Middlemost Brothers, Business Records

potential in the mineral reserves beneath its Huddersfield property and was anxious to develop these using sophisticated mining methods. To do so required finance and as with others developing collieries in the District this was to be undertaken without recourse to the capital market. Such landowners were used to managing agricultural estates where mineral extraction had been something of a by product. Changing emphasis from agriculture to direct involvement in mining to take advantage of the growing demand for coal presented a considerable risk for those who lacked expertise. Using one mining development to fund another may have been seen as a safer course than borrowing the money needed. The speculative nature of this venture is reflected In the fact that having acquired the leases the Thornhill estate then offered them as sub lets over the relatively short period of three years. This supports the view that the Thornhill estate was seeking a short term source of income to help fund developments of greater potential near Huddersfield. Local mill owners were quick to take advantage of these leases and the result was a rapid and intensive exploitation of the property. The first leases were granted in 1834 and in that year 2,800 tons of coal was extracted. By 1836 this had risen to 9,608 tons, rising again to 14,720 tons. A year later the rapid depletion of reserves was indicated by a fall to 3,415 tons and by 1837 it was clear that little remained with production falling to 1,415 tons[19]. All of the collieries on the site were simple pits and it is to be assumed that wasteful methods of mining were a feature. Teams of miners, usually about a dozen for each leasehold, contracted to work the coal for mill owners.

Through the wool textile industry it is possible to see how the locational advantages of local coal fuelled industrial expansion. This was not merely a question of proximity. The seams were often thin and of poor quality.

---

[19] Thornhill Papers

The ability to exploit these at low cost came via primitive mining techniques coupled with modest start up costs. These enabled mill owners often lacking technical knowledge, to exploit reserves close to their premises. If a mining venture failed then the financial loss might not be significant. If it succeeded then coal production could be under way in a matter of weeks. More sophisticated collieries would have been uneconomic. Indeed it is unlikely that mill owners would have had the finance or at least have been willing to undertake the risks associated with the sinking of a large colliery. The proximity of these coal reserves coupled with the ability to work them inexpensively meant mill owners were able to reduce manufacturing costs by up to a third[20] which was vital during the key decades of the 1840s and 50s when the railway network was still being developed. For the landowner the movement of the mills away from the traditional valley bottom sites meant that land previously of marginal value might now have considerable potential, not only for industrial development but also to provide space for housing. John Foster was an individual who had the various strands of the decision making process under his control. He was a landowner whose property included coal reserves. His background in both agriculture and domestic weaving enabled him to move forward into industrial weaving with a secure source of fuel. Elsewhere meeting the needs of the dominant industry for land and fuel was to provide a challenge for landowners.

---

[20] Musson A.E. I*ndustrial Motive Power in the United Kingdom 1800-1870*, Economic History Review 2nd series (1974) pp 419-420.

Chapter Four

THE ISOLATED MARKET

Whilst wool textiles was a major consumer of coal throughout the District, the two iron companies in Bradford had a major influence on both production and demand. Although Bowling and Low Moor were principally in the business of coal mining to meet their own needs, holding as they did such a dominant position in the local market meant that their presence in the period before the arrival of the railway was overwhelming. As the companies acquired more leaseholds so a network of tramways was developed to connect colliery operations to the Works. For example the Low Moor company was to acquire over 8,000 acres of mineral rights connected by 22 miles of tramways. This meant that any surplus of coal could be disposed of locally to either the domestic or industrial markets[1]. Bradford as a whole was to acquire 135 collieries and by the year 1856 was said to be producing over 1.8 million tons. This is a difficult figure to confirm and does not match the authoritative Hunt's *Mineral Statistics*[2]. However it does reveal how a vast internal market was created, much of this by the iron companies. The disposal of surplus coal onto the local market seems to have been less of a planned process and more a way of dealing with the effect of uncoordinated mining activity driven more by competition between the two companies for mineral rights, than the specific requirements of the Works. The Low Moor Ironworks had the strategic advantage of owning the Royd's Hall estate. In contrast the Bowling Ironworks had to look further

---

[1] Cudworth W., *Sketches of Low Moor and Bradford* (1876) p56
[2] Hunt R. *Mineral Statistics*

afield and some of its leaseholds were up to four miles distant. The small size of some leaseholds coupled with the often steep gradients presented Bowling with greater transportation problems than its rival. Clearly there was an incentive to run a tramway spur to a new colliery where the leasehold was of sufficient size to support mining for several years. Some leaseholds would never repay the investment and at best the coal mined had to be moved by wagon to the nearest loading point on the tramway network.

In Halifax the collieries on the Shibden Hall estate were well located to take advantage of the expanding market. This was because of the poor communications which made it difficult to transport coal over any distance before the arrival of the railway[3]. Initially the collieries had served the local domestic market and used its own idiosyncratic system for weighing coal. The 'load' was unique to the Shibden Hall collieries and seems to have evolved as a measure suitable for serving domestic customers. Other collieries used their own systems too. Often this was done on the basis of what a corve or basket could accommodate when being raised from the pit bottom. The diameter of the shaft then being the basis for the measurements sent perhaps to a local carpenter who had been commissioned to do the work. These local systems were common throughout the industry and had undergone a long period of evolution best documented in the North East of England. For example the Durham 'chaldron' measure was 20 hundredweight, the equivalent to one 'keel load' a measure used in the coastal shipping trade[4]. As long as the chaldron remained merely a measure it continued to increase in size only until the radical change from measure to weight was made and effected by statute in 1695[5]. The Newcastle chaldron was then declared to be its 'modern'

---

[3] Hartley W.P. *Five Landed Estates*, p194
[4] *Proceedings of the Archaeological Institute*, (Newcastle 1852) vol. I p169
[5] 6 &7 Will. III, c.10

weight of 53 cwt. The sheer complexity of weights used down the centuries even in a single locality, the north east, such as 'bolls' and 'wains' were aimed not only at the shipping trade but also the size of the vehicles used to transport coal from pit head to harbour. In the course of time this relationship was broken and began to be reshaped by the relationship between the lessee and landowner. This then evolved into the link between the size of the vessels used to raise the coal from the pit and the shaft diameter[6]. However the situation was further complicated by an additional relationship between producer and customer. This creates a problem where both producer and customer were dealing in an idiosyncratic measure such as a 'load'. Where records for both producer and customer (for example a woollen mill) still exist, each may refer to sale and receipt of loads without reference to a standard equivalent weight. How then may production and consumption be determined? The solution is as follows:

Colliery records usually state what *type* of coal was being mined and it is relatively simple using statistical sources such as *Hunt* to obtain the local price per ton of that coal. As a consequence if colliery accounts note that 140 loads of coal were sold at £11 2s 8d and the price of that type of coal was 4s per ton, then by dividing price into the amount sold reveals the standard equivalent weight to a 'load'.

---

[6] Galloway R.L. *Annals of Coal Mining and the Coal Trade* (London 1898) p36

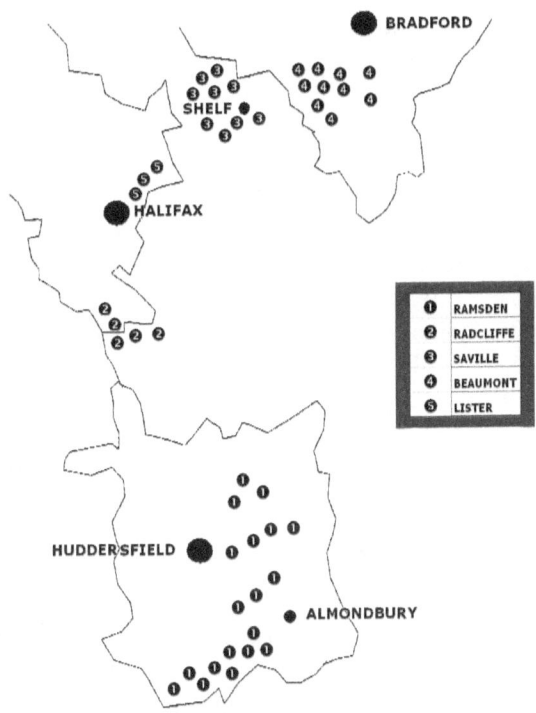

The Distribution of Collieries on Estates in West Yorkshire in 1850

The first industrial customer acquired by the Shibden Hall collieries was the Halifax Gas Company in 1849. This was a contract to supply 40,000 tons of coal per annum and marks the transition from the old style of mining on a landed estate; namely one of several income sources being derived from agriculture and other land based activities with a market among the local population[7]. Other customers were to follow such as Crossley and Leeming the carpet manufacturer in 1854. As a consequence the Shibden Hall collieries grew to be the largest local producers distributing mainly within a two mile radius.

---

[7] Hartley *Five Landed Estates* p195

The range of customers reflected the expanding industrial base of the town and included textiles, engineering and brass founding. This presents a good illustration of how a local landowner was prepared not only to undertake direct involvement in the coal industry but also to transform the collieries from primitive pits making local sales to the domestic market into an industrial supplier. In doing so the estate was taking a risk. It would have been possible to expand production by sinking more pits. Instead profits were used to invest in the equipment required to create a modern colliery enterprise. This was done at a time just before the railway had begun to establish itself in the town. The decision to do so may have been influenced by the sheer concentration of industrial customers in close proximity, creating good market opportunities that could be retained in face of competition from supplies brought in by the railway[8]. If this was the case then the Shibden Hall collieries were able to consolidate their position in the industrial market just before alternative sources of supply became available and as a consequence it was a risk that brought good rewards. Not only was the estate able to establish itself as a local supplier in the pre railway era but continued to serve this market throughout the greater part of the century. Names in colliery accounts from the later years reveal engineers, brick makers, dyers, schools, charities and of course various mills. In addition the estate began the process of finding niche markets, for example the manufacture of coke for the metallurgical trades[9].

The Shibden Hall collieries were not the only ones assisting industrial expansion. Within a circle of four miles 30 collieries were operating in 1852, the majority supplying engineering companies and the textiles trades[10]. As Langton has noted, because coal is heavy and bulky and because loss occurs in extremis during use, mining is

---

[8] Johnson J.F. *The Economy of a Coalfield* (London 1838) p120
[9] George E. *On the Yorkshire Coalfield* (London 1836) p174
[10] Report of the Select Committee on Coal Mines, Parliamentary Reports (1852) vol.1

likely to be closely tied to market locations[11]. Although broadly correct this argument overlooks two other factors. The first of these is the type of mining undertaken. As has previously been noted a necessary condition for rapid expansion to cater effectively for the industrial market was a primitive coal industry that could increase production not through investment in more sophisticated methods but rather by the simple expedient of increasing the number of pits. In this way the Lister collieries for example, were able to generate the revenue necessary to invest in more modern mining techniques and go on to retain their place in the market even as the railway network expanded. Secondly the complex assemblage of minerals enabled some coal masters to broaden the base of their income making it possible to work what might otherwise have been uneconomic coal seams. Arguably industrial expansion would have been much slower before the arrival of the railways without these factors being in existence.

Accompanying the expansion of industry was the increase in population. Nef[12] postulated the annual consumption of coal at between 9 and 16 cwt. per person or 45 to 80 cwt. per household. An alternative estimate was one ton per head of population[13].

Where a particular coal seam could not meet industrial requirements, domestic demand could provide an alternative market and is a further illustration of how when communications were poor seams of less than 18 inches considered by some modern authorities to have limited economic value were utilised[13].

The growth of the mining industry in the District was uncoordinated. Despite the bulk of the mineral reserves

---

[11] Langton J. *Geographical Change and Industrial Revolution* (C.U.P 2008) p42

[12] Nef, *op. cit.* pp 19-20

[13] Report of the Commissioners Appointed to Enquire into the Several Matters Relating to Coal in the United Kingdom, vol.1, general reports and sub reports, Parliamentary Reports (1871 vxiii reports commissioners)

[13] Jevons W.S. *The Coal Question* (London 1865) p129

lying in the hands of just a few landowners mainly belonging to the same small social class, no Lord Rhonnda emerged who could put forward a scheme to eliminate undue competition within the coalfield by means of a quota system[14]. The decision making process was undertaken by individuals with their own highly localised concerns. One reason for this and an important factor for consideration was noted by Court[15] who stated that:

'For many purposes it is a serious mistake to think of the coal industry as simply producing coal. Those who run mines think of themselves as supplying their customers with steam coal, household coal, gas coal or some other variety of coal that is in demand'.

Court might also have added fireclay, ironstone, or any other associated mineral.

In Huddersfield the market area for a particular colliery was often restricted by topography. This was a feature throughout the District where steep sided valleys made the movement of a bulky commodity difficult. In the pre railway era the solution was large numbers of collieries each serving a limited market.

Following Rostow the wool textiles industry took off in the decades 1770-1800. Self sustaining growth followed in 1800- 1830 and beyond came the drive to maturity which lasted for half a century. This mirrors almost exactly the expansion and ultimate decline of the coal mining industry in the District. Setting the consistency of a water supply against high fuel costs and technologically unreliable steam engines plus the high levels of maintenance often required to keep these engines in operation, it would at first glance have seemed wisest to retain water power in the Pennines. However not only was fuel cheap but as will be shown, the coal industry possessed flexibility sufficient to allow it to adapt to the needs of its major industrial

---

[14] Thomas D.A. *Some Notes on the Present State of the Coal Industry in the United Kingdom* (Cardiff 1896) p144

[15] Court C. *Some Problems of the State of the British Coal Industry* (London 1910) p214

customer. Its ability to do this in the pre railway era is owed in part to its agricultural origins. Further, as the century advanced there were improvements in the competitive position of steam power and the purchase price of new engines fell sharply. Continuous efforts were made to improve fuel economy.

The introduction of the factory system for cloth weaving saw water power predominant for several decades. Even before the start of the industrial revolution wool textiles production in Yorkshire had outstripped the rest of the country. This was despite the fact that only a few far sighted mill owners had chosen to install steam engines. The first mill in Yorkshire using steam powered frames had appeared at Addington on the River Wharfe in 1787. Soon afterwards came Meltham Mills near Huddersfield and Dean House Mill in Almondbury each in 1793, though in both cases this was done to augment water power. The introduction and distribution of Watt's steam engine was tied directly to the location of mill premises. On stronger streams a steam engine was not necessary but with the growth in the size of mills and the move away from valley bottom sites the use of steam increased. In Huddersfield space along river banks was at a premium or water power seen as often unreliable. A move to other sites could only be achieved in the pre railway era with a consistent supply of coal. By the 1830s there were 45 mills operating within the Huddersfield parish boundary and a further 42 in Almondbury with steam power predominant. The final improvements in mechanisation in the industry took place in the 1840s and 1850s. There was a heavy dependence on coal obtained locally, not only due to transport difficulties but also because the West Riding was unable to supply its own needs in the early years of the century. Indeed there was still an 'import' trade via Hull as late as 1829 and only after1850 did this alter to one of export when the more southerly coalfield around Doncaster came into operation.

Because of the policy adopted by the Ramsden estate

after 1840 the market for local coal in Huddersfield began to be distorted. Given the estate's hostility to mining close to the town centre, it became increasingly difficult for the remaining collieries in this area to maintain a market share. This was a significant development because no other estate in the District personifies the conflict between different types of land use to the same extent. The industry then was required to adapt early, before the competition from coal supplies brought in by rail began to have an impact. A pattern of adaptation was already evolving as local mines were forced to the periphery of the central area of Huddersfield.

Chapter three has charted the growth and expansion of the wool textiles industry: the increase in the domestic market can be gauged by the growth in house building. Between 1841 and 1851 a total of 5,955 new homes were added to the housing stock of the town each containing an average of 5.18 persons. This is reflected in activity on the Ramsden estate where site development during the years 1852 to 1898 showed continuing expansion except in the years 1878 and 1886[16]. The market for coal, both domestic and industrial had grown to such an extent that the loss of opportunities to exploit mineral rights in the central area did not have an impact. Mining continued on the peripherary and further afield. How it adapted to railway penetration and the further loss of markets will be considered in due course.

The Shibden Hall colliery accounts give an illustration of how this isolated market was served. Local hauliers and merchants would transport small quantities of coal to workshops or points of distribution to domestic customers. The pre industrial approach to coal sales was still working alongside supply to larger concerns. The year 1845 marks the beginning of a change. There was an increase in the labour force and the mines were now operating for fifty weeks per year. By 1855 the turning point was reached and

---

[16] Census Returns 1841-1901

customers were now exclusively some of the best known industrial names[17]. Given the dense concentration of industry in the Hebble valley the Shibden Hall collieries had a valuable market area even within a one mile radius of the estate. Whilst many other collieries in the area present the common picture of short working lives, the large area of coal bearing land, coupled with improved mining methods meant that the estate meant was able to consolidate and then hold its position as a major supplier to the local market. From the mid1840s with the arrival of the railway, the Shibden Hall collieries were faced with the same form of competition as those in Huddersfield.

Analysis of the market areas of small collieries is not easily undertaken since the identity and location of customers cannot always be determined. The smallest and most primitive collieries were those that served the local domestic market. Often this was done via sales at the pit head, sometimes called 'land sales'. Many small collieries did this and also sold to industrial customers, though in the main these were collieries capable of raising production to more than 3,000 tons per annum[18]. Whilst it is reasonable to suppose that even the smallest collieries were supplying small amounts of coal to workshops and other undertakings that could described as industrial; for the purposes of analysis the term is used where coal was supplied either to an engine house (to provide steam power to drive machinery) or for conversion into another product (as in the chemical industry).

Halifax's building cycle parallels that of Huddersfield during the years 1852- 1855. This was followed by a slump in 1858 with a recovery and rapid expansion during the 1860s. Whilst Halifax benefited from the early arrival of the railway in 1844, secondary transport links in the form of branch lines were slower in coming. As a consequence railway penetration of the market moved

---

[17] Shibden Hall Papers, CMA5
[18] Trigg W.B., *The Halifax Coalfield*, Transactions of the Halifax Antiquarian Society (1932) pp263-264

slowly. Halifax's population stood at 8,866 in 1801 and had risen to 50,600 by the end of the century, so there was a substantial domestic as well as industrial market[19]. The Shibden Hall estate had the locational advantage of lying in the valley bottom and had acquired a broad customer base before the arrival of the railway. Additionally it was rapidly acquiring the productive capability to meet the requirements of customers who might otherwise have looked to the railway.

To the north of Halifax the township of Northowram does yield some information on where the output of its three principal collieries was going in the year 1847. This is particularly useful since the influence of the Bradford iron companies was not making itself felt in Northowram at the time and as a consequence it is possible to see how the local mining industry was finding its own markets. Each colliery was producing in the region of 7-10,000 tons per annum. Northowram Colliery for example, was a series of shallow pits where the number was increased as necessary to cope with demand. New Road Colliery made the transition to deeper operations using a single shaft to raise the coal[20]. These were collieries expanding to meet demand from local mills and foundries. Another colliery New Bank, continued to use primitive methods of extraction and remained as a supplier solely to the domestic market throughout its working life. Ahead of competition from supplies brought in by the arrival of the railway each of the three collieries had found and secured a niche in the local market. Two were to evolve into exclusively industrial suppliers in a district where demand was rising due to the expansion of the wool textiles industry, whilst the third was carrying out the no less important task of supplying the domestic market.

In each case the market areas of the three collieries was within a one mile radius of the pit head, well within

---

[19] Glover *op. cit.*
[20] Shibden Hall Papers CMA5

Nef's[21] measure for an eighteenth century colliery and still applicable where moving supplies over a greater distance would have been difficult.

A similar picture can be found in the Holme Valley to the west of Huddersfield. Here in 1845 was another transitional coalfield originally established to serve the local domestic market. The arrival of the railway would not occur for another ten years and the distribution pattern of the collieries reflects both the isolation of the valley and the difficulties of transporting coal over any but short distances.

Essentially there are four clustering's of collieries to be found and each represents a pocket of growing industrial activity. The largest number of collieries, six in all was to be found unsurprisingly around the largest settlement Holmfirth[22]. The town acquired a gasworks as early as 1843, unusual for such a small settlement as yet unconnected to the railway. The availability of a local supply of coal is likely to have prompted this. Given the steep gradients of the valley it is unsurprising to discover that none of these collieries appears to have had a market area of more than a one mile radius. In part this would have been due to their supplying woollen mills already established along the line of the river which were converting to steam power. Also industrial development along the valley took place in pockets, constrained by the steep topography. There was not the continuous linear activity which occurred in the neighbouring but much broader Colne Valley.

The picture in Bradford is rather more complex and it must be remembered that the area under examination is Bradford Dale. Other parts of the city were also involved in coal mining but the significance of the Dale lies in the early concentration of industries, the landowners who leased or worked the mineral rights and the iron making and textile industries that required coal. The dominating

---

[21] Nef, *Ibid.*
[22] Wray *op. cit.* p154

influence exercised by the iron companies made Bradford Dale and the surrounding settlements different to mining activity carried out in the rest of the city. Apart from iron making and textiles markets for coal included engineering industries and lime burning.

Other landowners in the District were effectively constrained by the giant concerns at Low Moor and Bowling. Unlike much of the mining industry in the District the iron companies possessed technical expertise and this combined with its demand for coal and iron ore meant that the networks of wagon ways which were established extended their influence throughout the Dale and beyond. This was in contrast to the collieries of Huddersfield and Halifax where market areas were little different to those of the pre industrial era. In Bradford the iron works were able to extend the market area of more remote collieries and industrialise local settlements such as Wibsey. The Low Moor Company also sold coal on the domestic market, presumably encouraged to do so by the presence of poorer quality coal unsuitable for iron making. Such was the output of the company that some metallurgical grade coal was sent beyond the district. Interestingly this found its way to the port of Hull and was noted as the being greatest distance away of any colliery using the port. Low Moor was able to compete with coal of a similar quality from Newcastle[23].

The Bowling Ironworks also had a long reach. Local settlements were industrialised, starting with the point of origin at Bowling and moving on to Horton, Hipperholme, Tong and Thornton. Independent colliery owners found a ready market for their coal at the ironworks. Other markets included a gasworks and textiles mills. Smaller iron founders such as Thwaite Brothers and EW Haley were also operating by the 1850s but were becoming increasingly dependant on coal supplies brought in by rail. The population of Bradford reflected its rapid

---

[23] Brigg J. *The Industries and Geology of Bradford* (1879) p229

industrialisation. In 1801 it stood at 13,264 and had risen to 103,786 by 1851. The township of Bowling alone had risen from 2,055 in 1801 to 13,544 in 1851. Inevitably with the huge demand for coal from the ironworks railway penetration of the market from the 1840s onwards took place quickly. This is reflected by the fact that the number of coal dealers operating from railway sidings or canal wharves stood at 72 by 1875[24] and it may be assumed then that they were trading in supplies brought in from outside the District.

Operating conditions then, were far from uniform in the isolated market. The only common factor was that the increase in households and the expansion of industry fuelled by local supplies of coal. This was a phenomenon which would last for only a few years before the railways arrived. It is the domestic market which gives an indication of how marginal coal seams could still be viable; in effect achieving a great deal with very little. Where the coal was of poor quality or the seams thin then other minerals such as stone and fireclay could be worked and often used. Stone and even poor quality fireclay had their uses as low cost building materials that could be acquired close to the points where they were required.[25].

In the south of the District the Ramsden estate in forcing the coal industry from the central area of Huddersfield, provides an early example of how a market could be reshaped even in the pre railway era. The effect of this was to give colliery owners in the south of the District an early lesson in reacting to changing market conditions even before the pressure of railway penetration was felt. Further north the Shibden Hall collieries, directly operated by the estate, used their brief dominant position in the Halifax market to modernise. In doing so they secured a range of local customers and as a consequence were in a strong position as the railway supply network grew.

As an estate owner the Low Moor Company was well

---

[24] Baines Trades Directory (1875)
[25] Samuel R. *Miners Quarrymen and Salt workers* (London 1977) p18

placed for building a network of tramways to link collieries to the works. In contrast Bowling had to go further afield much sooner. Such was the power of the company that it was soon able to extend its influence into more distant settlements in order to lease mineral rights. Both companies were prepared to use whatever mining method was most suitable for a particular leasehold. By the 1840s the iron companies were operating mines capable of producing more than 10,000 tons per annum, working both coal and ironstone, along with primitive pits. Essentially it was the size of the leasehold that governed the choice of mining method. For the iron companies such was their dominance of the available leaseholds emphasised by the partnerships they created with the two main landowners; Beaumont with Bowling and Savile with Low Moor, that they had resolved their supply requirements by the 1850s for the next two decades. The loss of isolation brought by the arrival of the railway was to have no impact here. Instead supplies brought in by rail went to the textiles and other industries to augment the output of collieries not producing for the iron companies.

Several of the factors at work were in evidence wherever industrialisation was taking place: a dominant industry generally wool textiles which was rapidly adopting steam power, coupled with a growth in population and a rising demand for fuel. What gives the District its unique feature, certainly during the period 1838 to 1850 was its poor communications. There were other locations in the West Riding, notably Leeds and Wakefield better situated to accommodate an expanding wool textiles industry. Here not only was railway expansion more easily undertaken but a fast developing mining industry on a very large scale was able to meet demand. The isolated market with its thin coal seams was able to do so because the primitive mining methods proved to be sufficiently flexible.

This period also provides further evidence of the factors that had to be taken into account by landowners wishing to generate income from their properties. In Huddersfield the dominant position of the Ramsden estate not only caused mining to be driven to the fringes of the central area given the preference for building developments, it also interfered with mining operations on the Thornhill properties. Further north the Radcliffe estate permitted a destructive and inefficient form of mining to gain revenue whilst the opportunity still existed. In Halifax the Lister estate chose the course of direct involvement in mining and through this secured a market share that was able to resist subsequent competition from the railways. In Bradford Dale the iron companies via their respective partnerships with the Beaumont and Savile estates established a dominant position in land use. This provides the third and most complex element in the decision making process: to permit mining and if so in what way? Direct involvement or simply leasing the mineral rights with the attendant risks or indeed whether to join in a strategic partnership with a consumer.

Chapter 5

The Operation of a Small Colliery

For the majority of collieries in the District their working lives like the miners who toiled in them were carried out in obscurity. Where documentary evidence still exists such records can be fragmentary and a picture usually has to be built up from a variety of sources. This is particularly the case for collieries established in the first half of the century, working leasehold on one of the landed estates and seeking a market among the local mills and households. An exception to this where the records have been unusually well preserved is that of Springwood Colliery. As a result it has been chosen to illustrate in detail the type of operation that underpinned the industry and its productivity. Surface activity at such collieries often reflected the temporary nature of the operation. Occasionally a longer lease would include a clause permitting the removal of stone for the construction of buildings but generally little effort was expended on this. Temporary buildings were the norm.

The Springwood estate lay to the west of the main railway station in Huddersfield and in 1861 a preliminary survey was carried out to determine whether coal existed in workable quantities. The surveyor reported to the estate Steward that there were 15 acres of coal to be worked and that 36,300 tons could be extracted. This would have been an estimate based on the assumption that the seam would run uniformly beneath the property. The surveyor advised leaving a third of the coal untouched. The net quantity then was 24,200 tons[1]. This suggests the intention was to adopt the 'pillar and stall' method of working, which whilst not

---

[1] Ramsden Papers RA8

the most economic approach did have advantages in a shallow coalfield in that it reduced the risk of subsidence and surface damage.

The records of Listerwick Colliery on the Shibden Hall estate provide details of the operation before it had evolved into a single site operation. During the 1840s the colliery employed between 15 and 20 miners. Production averaged 1,000 loads per week, (40 loads equalled four tons at this colliery) Gifts were made to colliers at midsummer, suggesting that agricultural customs on the estate were extended to those working underground. Total expenses for the half year in 1846 amounted to £101 18s illustrating the modest cost of operating a primitive colliery. In that year the average wage for an overman in the industry was 5s and for a miner 3s 10p [2]. Collieries operated by John Foster owner of the Black Dyke Mill reflect the same type of operation; a series of pits with much coal left unworked to support buildings. Without the assistance of mining engineers it is not surprising to discover frequent references to surface damage caused by this type of mining[3]. The process of getting a colliery into operation could be achieved quickly. One pit on Foster's property was sunk in six days at a cost of £14 6s 1d. Following this the actual mining of coal might be undertaken cautiously. Foster's New Keelham Colliery which opened in 1846 began by employing only two miners, averaging between 8 and 15 loads per day until Foster was satisfied that the coal existed in sufficient quantities.

Where a small colliery was directly operated by a landowner the miners were employed in the same way as other workers. Wage rates reflected the norm in the locality but there was little organisation. The miners at a particular colliery would deal with the estate steward. There appears to have been no use of agents to negotiate wages and conditions, the precursor to union

---

[2] Shibden Hall Papers CMA6
[3] John Foster, Business Records

representation in some parts of the country[4], nor was it generally the practice to employ young children. Inspection of such operations was haphazard. Visits by the Mines Inspectorate took place only on request or as the result of a complaint[5]. It was not surprising then that individuals with no experience of mining but possession of a few acres of land beneath which lay coal reserves, were prepared to become involved in the trade. For example at Heckmondwike in 1850 a grocer by the name of Grearley rented fields which he covered in day holes, mining the coal for sale to local householders. For deeper mines the usual method was longwall working, leaving pillars in place. This was a method which had originated in Shropshire during the 18th century. The first references to its use in the District occur in the 1820s where it was quickly adopted, being an effective if uneconomic method of working in shallower parts of the coalfield[6]. One benefit of working the thin seams was that the risk of gas and explosions was low, meaning that the knowledge required to deal with a major underground hazard was not always required. Miners were recruited from the general population and apart from those employed by the Bradford ironworks there were no mining communities as such. The development of a coalfield might see many of the local population employed in mining, though depending upon the duration of the leasehold and the amount of reserves to be worked it could easily be a temporary phenomenon quite unlike the mining communities developed further east in the county. It was common then for those employed in mining to be part of the general working population and on landed estates miners might also double as agricultural workers.

---

[4] Report of a Select Committee Appointed to Enquire into the Acts for the Operation and Inspection of Mines (1866) vol xiv, Reports Commissioners (9).
[5] Samuel *op. cit.* p19
[6] Ashton & Sykes, *The Coal Industry of the Eighteenth Century* (London 1924) p30

For anyone with the modest capital required to enter the mining industry there was a fixed rent to pay for each acre of coal bearing land and a royalty payable on the amount of coal produced. When output fell so that royalty was less than the fixed rent this was known as 'shorts' and the lessee was sometimes able to deduct it from the sums payable for output in excess of the minimum in subsequent years[7]. In this way the small colliery could stretch its operating costs and remain viable over periods when demand was low. Leases were generally derived from agricultural practice; in effect more a form of sale than an actual lease which may be interpreted more as purchase than rent. The lessee was in effect tendering for a mineral field, taking into consideration the quality of the coal, the cost of production and the cost of carrying away afterwards. This generally included all wayleaves after which the lessee then makes up the royalty which he believes he can afford to pay. Royalty is reserved by way of a proportion of the sale price generally, described as the sliding scale by way of a fixed sum per unit, ie per ton/per unit. Alternatives might be by way of a proportion of the net profits or a fixed annual sum irrespective of the quantity worked[8]. The advantage of primitive methods though essentially wasteful, was that during the early phase when the market grew in isolation, there was little reduction in production due to the difficulty and cost of expanding older collieries, where there might not have been further leaseholds to be had in close proximity. This contradicts the view set out by Rostow[9] who noted that the effects of an early start are acutely felt because the layout of a mine at any given time is partly a reflection of the state of mining technology when it was first sunk; the older the productive units in the industry the less efficient

---

[7] James C.A. *Mining Royalties, Their Practical Operation and Effect* (London 1898) pp20&67
[8] James *Ibid.*
[9] Rostow W.W. *The Process of Economic Growth* (London 1960) pp101-103

they are likely to be. Whilst this would be the case in a deep mine with extensive workings, it would not apply to small primitive collieries. The small owner working shallow seams whose capital resources are limited is perfectly well able to compete with large deep well equipped mines. His ability to do so is a test of his efficiency. Productivity (output per man) is an imperfect measure of efficiency. The real test of efficiency is the extent to which actual performance measures up to potential performance under particular physical circumstances[10]. This comment effectively encapsulates the operation of small mines in the District. The use of steam power for raising coal to the surface, an obvious way to support an increase in production, was an unnecessary expense when a number of pits were being worked simultaneously. Horse gins were the favoured method; a simple construction consisting of a light drum of wood or iron about 18 inches deep and 7-12 feet in diameter. This was fixed above the shaft supported by a wooden framework attached to a headframe. A strong wooden pole was attached to the drum and a horse harnessed to this. Hemp ropes of about one and a quarter inch diameter were used for winding the corves up the shaft[11].

At Springwood Colliery the coal extracted would have a value of 6s per ton and this would realise £7,260. The cost of working the coal was estimated at 2s 6d per ton, leaving a profit of 3s 6d. Therefore the net value of 24,200 tons would be £3,025. The colliery would require a 20 yard shaft costing £20 to sink. Springwood colliery was not it seemed to have the luxury of twin shafts, an important safety requirement which subsequently became a legal obligation on colliery owners. This was in fact generally the case with Huddersfield mines at that time. The nearby New Ground colliery was said to be one of the

---

[10] Alcroft DH., *The British Coal Mining Industry* (1974) p109

[11] Cudworth W., *The Histories of Boulton & Bowling* (Bradford 1891) p42

few with twin shafts.

Even as late as the 1850s the Ramsden Estate did not choose to seek the services of a professional mining engineer to survey the colliery; the estate surveyor being required to combine his routine duties with this task. Significantly his surviving weekly reports[12] show that he was finding it necessary to devote an increasing amount of time to this one task. It was to be a further ten years before the estate used a professional mining engineer.

The final surveys were completed towards the end of 1861 and by 1862 work had begun on the site. No record remains of how long the shaft sinking took or what the dimensions were. New Ground Colliery had oval shaped shafts with dimensions of 9 x 8 feet and as the scale of operations at both mines was similar it is probable that the Springwood Colliery shaft was the same size.

The usual horse gin was used for raising the spoil when the shaft was sunk and this method was temporarily retained for working once the colliery became operational. Subsequently however a more robust head frame was erected for winding, this being constructed from timber. Few estate based mines used costly iron head frames and buildings of brick or stone were also a rarity. Springwood colliery though was eventually to get an engine house built.

By late 1862 the colliery was ready to begin working. As with the rural areas of the District, Huddersfield did not have an established mining community from which to acquire a workforce. An examination of the census data covering the inner wards of the town reveals only a small number of men who described themselves as miners[13]. Whilst there are examples of villages where mining evolved to become the single most important activity, the wool textiles industry maintained its position as the dominant employer activity. Only in those locations close to the Bradford ironworks where the companies often

---

[12] Ramsden Estate Papers R.A.4
[13] OPCS, census data (1861) Huddersfield inner wards

provided housing could significant clustering's of miners as a distinct occupational group be found. Elsewhere, given the short working lives of the collieries mining might be a transitory occupation, essentially a better paid form of labouring.

The number of miners employed at Springwood Colliery varied considerably. In the early months when production was getting underway between twelve and fifteen men plus several boys were working underground. After development was complete the number of miners fell to six. On the surface a further three men were working, chiefly on winding duties and moving the coal. The view of mining as something less than a specific trade is supported in correspondence written by the man brought in to manage the mine. He noted that whilst working in the wool trade could be arduous, conditions were far better than those to be found in a shallow, poorly equipped mine. The manager's reports regularly cite a return to the textiles trade as a miner's reason for leaving[14].

From the outset conditions in the colliery were poor. The shallow Huddersfield mines were often wet and Springwood Colliery was no exception. It was unusual for steam power to be installed in a small colliery but at Springwood this was felt to be unavoidable and the manager set aside a proportion of the weekly output to run drainage pumps. The obvious alternative to this would have been to dig a sough or drainage tunnel and allow gravity to do the work. However this was a mine operating on the fringes of the urban area. The scope for further tunnelling beyond the site was probably limited by building projects which the estate would not wish to jeopardise. The cost of this coal for running the pumps could be £3 per week which meant that over twelve tons of 'average engine' as Springwood's coal was described, went back into the running of the mine. There was then the

---

[14] Hartley W.P. *Springwood Colliery Huddersfield: A Portrait of a Yorkshire Estate Coal Mine 1862-77* Yorkshire Archaeological Journal, vol. 53 (1981) p94

paradox of steam power being overlooked in favour of a cheaper less efficient option for raising the coal but still being required to keep the mine in operation.

As was generally the case in the District one underground hazard that the miners of Springwood Colliery did not have to contend with was gas. This probably contributed to the generally low casualty rate in Huddersfield mines. In fact throughout the entire history of mining in Huddersfield during the nineteenth century only one major accident involving gas was reported. The Mines Inspectorate did however report two fatal accidents at Springwood and the mine manager was regularly in the business of seeking replacements for miners who had suffered broken bones and other injuries incurred underground[15].

In 1871 after various problems mostly geological in nature which resulted in a rapidly falling output, an independent mine viewer was called in. In a long letter which comprised his report he wrote of 'poor working conditions' and 'inadequate ventilation' plus non compliance with 'government regulations'. He also considered that the manager was not paid enough, a hint perhaps that he lacked the expertise to run the mine effectively and that he had little incentive to improve output. Springwood Colliery was certainly not alone in this. The nearby Fieldhouse Colliery had been reported by the Mines Inspectorate for employing under age boys and another Huddersfield colliery Taylor Hill, for being badly ventilated and a cause of subsidence in the locality.

Some of the report's recommendations appear to have been acted upon. A new air shaft was sunk later in the year and the pumping machinery was overhauled, which resulted in a fall in coal consumption by the engine. The overall result in 1872 was an improvement in output of 70 tons per week. Despite its inefficiencies Springwood Colliery was a viable business assisted by the fact that its

---

[15] HM Mines Inspectorate Reports (1860-61)

main customer Newtown Mills lay close by. This meant that little was required by way of facilities to transport coal from pit head to customer, except for an improved road suitable for moving heavily laden wagons. A comparison between the engine house records of the mill and colliery production, shows that the mine was in the main able to supply most of the monthly fuel requirements [16]. Occasionally supplies were sought from other sources but the explanation given was that a great deal of overtime was being worked at the mill. Equally the colliery would sometimes find itself with a surplus of coal. There was no system of holding any in reserve which may have been due to a lack of space for storage or equally a refusal to tie up capital. On such occasions the coal was disposed of on the local domestic market.

In 1874 the colliery began to have problems once again. A roof fall in May of that year halted production for a week and extra help had to be called in to clear the blockage. A series of letters between mine manager and estate surveyor included some bitter exchanges as to where the blame lay. The surveyor was accused by the manager of supplying poor quality timbers for roof supports. This sounds rather like a man who is likely to have to answer for a week's loss of production. Presumably he expected little understanding if his explanation was geological in origin and sought to move the blame elsewhere.

Two months later the pumps were out of action because of a breakdown. Subsequent reports unsurprisingly speak of very wet conditions underground; a situation that prevailed for some weeks even after they were back in operation. Production fell because of this and Newtown Mill was forced to look elsewhere for supplies. In August of that year there were complaints to the mine manager that even though the mill had been closed for a holiday coal stocks were still low. At the end of 1874 another colliery viewer examined the mine. Confirming his

---

[16] Huddersfield Tolson Museum, Newtown Mill Accounts

predecessor's reports of poor conditions, he predicted that the mine would be exhausted in 'three or four year's time'[17].

Considering the modest amount of money invested in start up and running costs, the mine brought the owners a good return. For example although the coal was originally valued at 6s per ton the actual value was 5s 8d . Despite this the overall profit came to £4,370 which was £1,345 more than predicted. It must be assumed that these profits were realised via lower than anticipated operating costs.

With the exception of the use of steam power for drainage Springwood Colliery provides a good example of the type of colliery undertaking that typified coal mining in the District. It operated on a modest acreage of 15 acres (although fewer than five acres might be worked). The mine was shallow and poorly equipped. Underground conditions were primitive and output was adjusted to the demands of a single customer. Professional expertise was not used in the development of the mine and its shortcomings were only exposed years later. The working life of about thirteen years was good under the circumstances and given the encroachment of other types of land use on the fringes of an expanding urban area, it is unlikely that scope would have existed to extend the mine any further. This would of course have acted as a disincentive to improvements in equipment or infrastructure.

The sinking and subsequent operation of Springwood Colliery provides a rare example of the entire working life of a small colliery working a leasehold on one of the landed estates in the District. From this source it is possible to identify some of the key factors in the decision making process. As has been shown a small colliery working a modest leasehold in a shallow part of the coalfield could be brought into operation quite quickly and cheaply, to generate a source of income for the estate.

---

[17] Ramsden Papers RA4

Springwood Colliery was though very close to the perimeter of the central area where the estate had some years previously forbidden any further mining development in order to preserve the land for building development. The arguments in favour of permitting mining in this location would have been made by people with a very limited knowledge and therefore the risk factor was considerable. The ability of a small colliery to be established on a lease close to its major customer would however have been a considerable incentive to proceed. It may also have been the case that the railway line which crossed the town west to east was seen as a barrier to any significant building developments and consequently mining was unlikely to lower the value of the land. In fact this proved to be the case and in the latter part of the century the area was used for low value terraced housing.

Chapter Six

The Transportation of Coal

Broadly speaking questions regarding the movement of coal from pit head to customer may be divided into two periods; pre railway and beyond. The picture is made more complicated though by the often steep topography of the District which at times served as a barrier to effective communications. Coal is of course a bulk commodity and moving it any distance can be laborious if only muscle power is available. Prior to the arrival of the railway parts of the District were served by canal. Indeed the construction of the Ramsden canal served to link the Huddersfield Narrow (better known as the Rochdale canal) with the Calder and Hebble Navigation. This provided access to coal supplies from both sides of the Pennines. The canal network had its limitations. Pennine winters could act as a handicap to movement of goods on water and as a means of transport it was slow and its trading hinterland was limited by the steep topography. More remote settlements could not rely on the canal as means of obtaining coal.

In terms of the larger conurbations Bradford had no decisive advantage in the matter of canal transport after 1777. By this time the city was linked with Leeds and hence to the Leeds-Liverpool canal. The River Calder had been made navigable as far as Sowerby Bridge as early as 1758 and in 1828 Halifax was linked to the canal network[1].

As has been noted the market area for many small collieries continued to fit Nef's estimate for eighteenth century collieries. Only the Bradford iron companies

---

[1] Boughey J., *Hadfield's British Canals* (1998) p193

developed tramways for the movement of coal. This was undertaken out of necessity. Such was the demand for coal and ironstone that the companies were forced to reach out further for supplies well before the arrival of the railways. Elsewhere in the District the model set out by Taafe[2] for the diffusion of internal transport lines suggest an explanation. Although Taafe based his theory on a study of growth in a colonial setting the model is useful in explaining the expansion of the District's railway network. *Phase one* in Taafe's model refers to coastal settlements but can be adapted to take into account the construction of the two major trans Pennine routes between Manchester and Leeds which was connected to Huddersfield in 1849. Halifax was connected to the railway in 1844 and Bradford 1846. *Phase two* consists of the emergence of a few major lines of penetration, for example routes linking Bradford to Halifax in 1850. The significance of these developments can be measured by the increase in the number of coal merchants operating from railway sidings and limited market area of local collieries. The arrival of the railway began the process of coalescing almost the entire District into a single market, providing coal merchants with convenient distribution points were to fill gaps where local producers could not provide supplies in sufficient quantity. The growth in the number of coal merchants operating from railway sidings in the central area of Huddersfield during the 1850s provides a good example. Here unlike other parts of the borough there were no collieries operating due to the policy of the Ramsden estate to restrict mining, consequently some local collieries supplying customers within this location were at the limits of their market areas. In contrast the railway was able to penetrate the central area and the market was then quickly dominated by coal merchants utilising this source of coal. *Phase three* consists of the growth of feeder routes and the beginnings of lateral interconnection. An example of this

---

[2] Haggett P. & Arnold E. *Locational Analysis in Human Geography* (London 1969) pp80-81

would be the building in 1852 of a short section of railway linking Huddersfield and Halifax. *Phase four* repeats the process of linkage and concentration. Within the District this can be interpreted as the final connections which saw the completion of railway penetration to all parts by the 1860s. One of the first branch lines reached Wibsey as early as 1848 and then ran on to Low Moor[3]. It is unlikely that either of the Bradford ironworks required access to coal supplies from further away at that stage of their development, a view supported by the fact that colliery numbers were still growing and further leaseholds were yet to be acquired. In general landowners in the district were not great promoters of railways. Their involvement tended to be more as facilitators, for example the Ramsden estate selling land that a railway company required in order to build a station in Huddersfield[4]. Bradford being a centre of the wool trade was an obvious place for the railways to come. The city was to occupy a central position for several rail routes by the 1850s and had acquired its first connection in 1846. Some idea of the demand for coal is given by the fact that in 1875 there were 100 coal dealers in Bradford of whom 92 had addresses at railway stations or sidings[5]. This number of dealers who can be assumed to have been wholly or partly dealing in coal brought by rail to the city strongly suggests that although Bradford was one of the most significant producers of coal in Yorkshire, output was not enough to supply the needs of the city. Arguably the need for supplies from other coalfields would have been much less but for the demands of the iron companies who were absorbing the bulk of production in Bradford Dale.

The surge in industrial output coupled with the expansion in population and the physical barriers common to the district meant that the local mining industry had two

---

[3] Cartwright J. *Illustrated History of Low Moor* (Bradford 1906) p77
[4] Ward J.T. *West Riding Landowners and the Railways* Journal of Transport History volume 4 (1960)
[5] White's *Trades Directory* (1875) p766

choices to stay in touch with its markets. The first was to develop an 'internal' transportation network, achieved only by the iron companies who possessed both the need to do so and the capital required. The second was to relocate, this is an option that will be discussed separately.

There was a further factor exacerbating the impact of railway penetration. It might be assumed that a branch line would be built for obvious reasons: the movement of goods and passengers in and out of settlements, which forms one of the basic assumptions of Taafe's model. Competition among railway companies has often been cited as the reason for duplication of some routes. However as the railway companies grew in wealth and influence another factor not previously explored came into play. Close to Huddersfield lies the small textiles town of Dewsbury which provides an example of this phenomenon. By 1861 Dewsbury's population had risen to 24,000 and the urban area had begun to consolidate. The town had in fact reached the limits of its prosperity and from the 1860s onwards a slow process of decline began[6]. Few new mills were opened after 1858. Like Huddersfield and Halifax, Dewsbury had a coal mining industry but it was inadequate to meet local needs. The first railway line to reach the town was the Manchester to Leeds line which passed through Huddersfield then on to Dewsbury in 1846. The line ran through the north of the town which was then the core of the textiles district. During the latter part of the decade the company (which was to become the London and North Western Railway) considered expanding its facilities and even went so far as to promote a private Act of Parliament for this purpose. Subsequently the idea was abandoned, presumably because the company noted the encroachment of other lines and the reduction in its market area. Logically this should have been part of the test for the development of any new railway enterprise. However

---

[6] Glover F.J. *The Rise of the Heavy Woollen Trade of the West Riding of Yorkshire in the Nineteenth Century* Business History, vols. 4-6 (1961-64)

fifteen years later to the south of the town on the southern bank of the River Calder, the rival Manchester to Leeds line of the Lancashire and Yorkshire Railway created a spur which terminated in a goods station. Although in construction terms this was a simple process, in business terms it made little sense. Ostensibly its purpose was to carry coal into the town and yet the location meant that the railway could only have access to a small portion of the market. In addition because the station was a terminus and operated on a small site, the space for coal trains was limited and this affected the ability of the company to run a complimentary outward flow of other types of traffic[7].

A third line then reached the town in 1864. This was a branch line running east to west from the Great Northern Railway's main Leeds to London line. Its ostensible purpose was the creation of a relief route around Leeds to Bradford[8]. At Earlsheaton a junction was created, which allowed the line to branch off and enter the Calder Valley via a tunnel. The business case for the creation of a third route into a town where the textiles trade had at best ceased to expand is questionable. Even stranger was the decision of the L&Y to enter into a partnership with the GNR, crossing the river via an expensive to construct viaduct and then creating a large terminus. Commercially this partnership seems to have made little sense. The existing L&Y goods station was not working to full capacity and by 1868 the number of mills in the town had fallen to 28.

In 1891 a fourth company completed a line to Dewsbury. Previously unrepresented in the area, the Midland Railway faced the problem at this late stage in the development of railways of finding a suitable route. The most obvious approach along the floor of the Calder Valley was already occupied by the L&Y and no space existed for a parallel line. Instead the company was forced to carry out the difficult and costly civil engineering work

---

[7] Marshal J. *The Lancashire and Yorkshire Railway* (1972) vol.2 p234
[8] Grinling R. *A History of the Great Northern Railway* (1975) p188

involved in building a railway along the valley side. This involved the construction of numerous high embankments and cuttings, plus two large and impressive viaducts. Finally where the route crossed the valley it was necessary to bridge all of the existing lines of communication which ran east to west along the valley floor. Upon reaching the town the line terminated at a goods station.

Working on the basis that a railway line must end somewhere an obvious means of enquiry is to consider the opportunities for traffic. Since the town's major industry had begun a slow decline and therefore the coal trade was unlikely to increase, there is the question of business along the route. Both the GNR and Midland lines show a paucity of both settlement and industry that might have prompted these extensions of their network. The original Lancashire and Yorkshire station had been created where there was limited scope for business. Why then was this expansion taking place and was it replicated elsewhere?

It is suggested that there were two reasons for this. The first being prestige. When an organisation such as a railway company is performing well and generating profits then the internal dynamics may require another outlet. Free of the criticism that would come if the company was underperforming, energy is directed elsewhere. In the case of these railway companies it was 'invading' the territory already occupied by a rival even if it made no commercial sense to do so[9].

The second reason was strategic; the aim being either to occupy an empty space on the map before another company did so, or in order to forge an alliance with or gain concessions from a rival. For example a clue to the Midland Railway's real intention when entering this already overcrowded market is to be found in a speech by the company's chairman. *"It has been urged upon us for many years that the Midland Company should be*

---

[9] Hartley W.P. *Some Factors in Railway Route Convergence : A West Yorkshire Example* Journal of Regional and Local Studies vol18 No2 (1998)

physically represented in the district which it is not at present.....that we should come and do our business first hand instead of at second hand as we do now..."[8]

By *appearing* to do so the Midland was able to convince the L&Y that it would be wiser to grant concessions in the shape of running powers over its lines. In this way the company gained access to Huddersfield and beyond. This was not an isolated example. In 1850 the L&Y ran a branch line off its Huddersfield to Sheffield route whose ostensible aim was to reach Barnsley. The line though came to a halt in the village of Scissett to the south of Huddersfield. There was no business case for linking the Barnsley coalfield to Huddersfield. Penetration of the Huddersfield market by the L&Y was effectively complete by this time, since the company's branch lines had reached all of the nearby valleys. Additionally the L&Y already had access to coalfields on both sides of the Pennines. In fact the line came to a halt having achieved its aim. This was to place a route barrier across a section of countryside where no other railway existed. This part of the map was in effect a void to be filled by a strategic move which was to deny any other company use of the route. Given the topography of the coal measures countryside alternatives were often unavailable. For the coal industry of the District railway penetration might not then necessarily follow business logic and was therefore difficult to predict.

---

[8] *The Railway News* (March 1890)

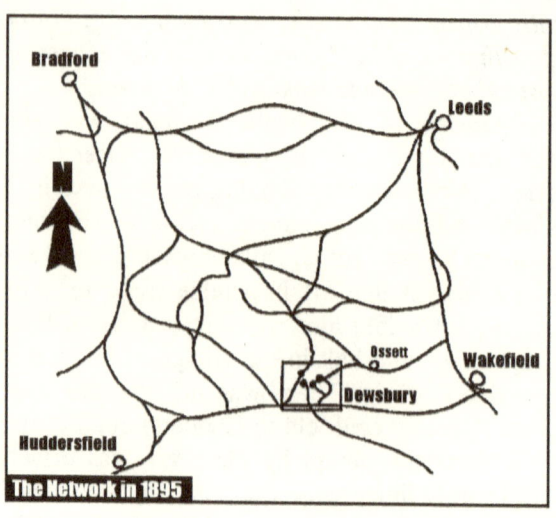

There was a progressive spread of mine workings in Bradford Dale which had begun in the seventeenth and eighteenth century. This was to include such settlements as Bowling, Bierley, Wibsey, Shelf and Tong among others in a semi circle around the southern flanks of the city. Such were the number of sites that there was an obvious incentive to create a system of tramways. In order to send coal to the ironworks the Low Moor company had by the end of the century a network covering 22 miles and these extended over an area of 8,000 acres. The Bowling Ironworks used a system of narrow gauge lines to transport the coal. This consisted of 21 inch and 24 inch track[10]. These integrated networks were the only ones of their kind in the District. In Bradford Dale then, there was the impact of two large industrial concerns creating their own internal transportation system which played an important part in ensuring that their demands for coal could be met locally. In Huddersfield despite the fact that the local coal industry would have been assisted by something similar but more

---

[10] Cudworth W., *Sketches of Low Moor and Bowling* Bradford (1876) p114

modest in scope, the Ramsden estate used its dominant position to prevent the development.

The picture in Halifax was somewhat different. The railway had arrived in 1844 and here there was no dominant landowner. The Lister family of Shibden Hall were the owners of the largest block of mineral rights within the District and worked these directly. Because the collieries had been modernized their productive capacity was increased and the collieries were able to gain and hold a market in Halifax despite competition from the railways. This was in part due to the estate negotiating long term contracts something that was beyond other coal masters in the District due to the small size of their collieries.

The arrival of a main line railway in Halifax had a similar impact in the central area to that in Huddersfield . However this was because the main concentration of industry remained in the bottom of the Hebble valley. Although expansion onto green field sites did occur it was much less pronounced than in Huddersfield due to the steeper gradients. Smaller collieries close to the central area quickly yielded markets to railway competition but because of the steep topography those serving locations remote from the railway were protected for much longer than those in Huddersfield.

Isolation might provide only temporary protection. The township of Meltham west of Huddersfield illustrates how quickly the arrival of a railway could effectively destroy a local mining industry. Meltham was an isolated location at the head of a small valley with a mining industry based on a coal seam two feet in thickness. Following a familiar pattern this coal had been worked initially for the domestic market but by the 1840s had expanded to serve industrial customers, both woollen mills and local brickworks. A branch line from Huddersfield arrived in 1850 and within five weeks the local coal industry had ceased production. There was no hinterland beyond Meltham where the industry could seek new markets so if the local collieries could not retain customers within the township then failure was inevitable[11].

The transportation of coal beyond very limited market areas began with canals. Huddersfield, Halifax and Bradford all benefited from canal links that were established in the eighteenth century but these had only a limited impact on the local market. They did though play a useful role in stimulating industrial expansion and the gradual move to steam power by the mills. Their limitations; lack of speed and seasonal failures limited the benefits of coal brought from distant sources and provided no significant competitive threat to the local industry. As the railway began to penetrate so there was no collective responsive to this far more effective threat to local producers. How the mining industry met and adapted to this competition will be considered in a subsequent chapter. In Huddersfield the dominant position of one local landowner was enough to frustrate attempts to move coal internally across the market. For Halifax topography was a barrier and the Bradford Iron companies were able to use their power and position to build tramways which extended the market area of many local collieries.

The opening up of the District to supplies brought in from distant coalfields brought about major changes which were to affect the local coal mining industry for the rest of its existence. For the landowner this added yet another factor in the decision making process. Whilst the business case for a railway might be quite evident it has been possible to show that commercial logic or even the desire to compete for traffic with a rival company were not the only factors at work. Railway companies might be motivated by something more than commercial imperatives and with the finance available were prepared to build lines simply to establish a presence. For the owners of mineral rights evaluating the business case for exploitation and taking decisions about the likely market for coal might not be enough. The prospects of railway

---

[11] Haigh A., *Railways in West Yorkshire* (1974) p14

penetration following an unpredictable route was a risk and might therefore encourage a landowner to permit primitive and inherently wasteful methods of extraction in order to exploit the coalfield whilst the opportunity still existed.

Chapter Seven

The Colliery Distribution Pattern in 1840

The year 1840 provides a useful point in time to make an examination of the colliery distribution pattern. At this stage the District still fitted the Isolated Market description discussed in chapter five, since no railway penetration had taken place. The starting point for this examination is to first consider the mining system devised by Losch[1]. Figure 1 below sets this out. In the 'Resource' section are the points already touched upon in previous chapters. The most notable of these is 'site' since this encompasses so much of the decision making process facing landowners. It will be remembered that the seam to be worked was often more complex than the Loschian model allows for, certainly in the District where a thin seam of coal of little viability in itself could be enhanced by close association with another mineral such as fireclay. The easy availability of labour and the modest start up costs often associated with developing a colliery in the District meant then that site was the most important element under the Resource heading, since there might be other uses for a particular location and the decision to mine could commit the landowner to a course of action which left little room for alternatives.

---

[1] Losch A., *The Economics of Location* (1940)

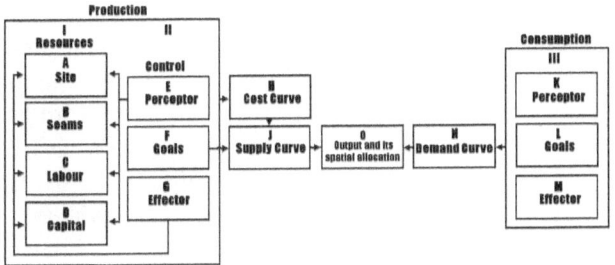

The Loschian model takes the enquiry only a short distance. Of greater value is the model devised by Langton[2]. This model encompasses more elements, some of which would have had no impact in the District, such as 'other fuels' and 'export'. Even transport has only limited significance in the pre railway era, since collieries operated in a market area largely unchanged in scope since the eighteenth century. However Langton's model takes on greater significance by the inclusion of the depth, quantity and quality of the seams to be worked, linking this to the competitive strategy. This then is a model which points towards the core of the decision making process and is allied more closely to the themes which have been covered in earlier chapters. Langton's model though has its limitations in the sense that the variables do not encompass all of the factors at work in a closed market. For example whilst it may appear presumptuous to dismiss 'Transport' as only a minor consideration, this needs to be seen in the context of a largely primitive mining industry and the reason for doing so will become more apparent when the subsequent shape of the colliery distribution pattern is considered in a later chapter. In fact it may be better recall the closing remarks of the previous chapter and consider Transport in the context of a further risk factor for landowners, in the sense that railway penetration could

---

[2] Langton J., *Geographical Change and Industrial Revolution* (2008)

affect the viability of a marginal coalfield. The decision to build a railway was not necessarily based on just a business case. A strategic addition to the railway map or the desire to enter the territory of a rival might be the sort of variables that could affect a landowner's decision making. However with no railway penetration at this time the industry was expanding both in colliery numbers and output to meet the demands of an expanding population and industrialisation. Although railways were absent, they were a factor that landowners would have been aware of[3]. Indeed some landowners were actively agitating for rail links [4].

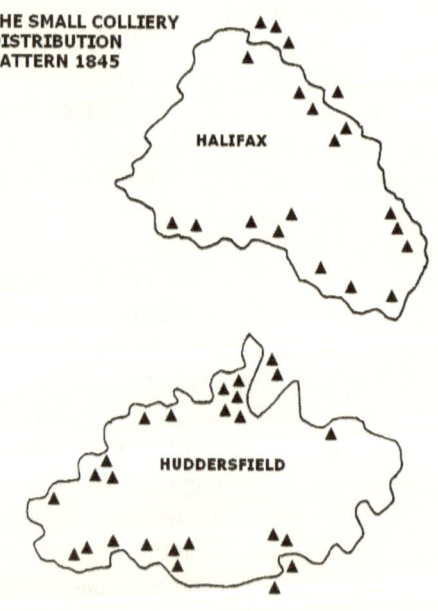

By 1840 the effects of the Ramsden estate's policy left only two small collieries still operating close to the central

---

[3] Ward J.T. 'West Riding Landowners and the Railways' *Journal of Transport History* (1960)

[4] Ward *op. cit.*

area of the town. Both of these were insignificant in terms of output and it may be assumed that these were collieries of the most primitive kind, probably working a series of shallow pits and exactly the kind of operation that the estate was anxious to eliminate from its properties. Along the southern boundary of the borough lay a close concentration of collieries. This reflects the different operating conditions on the Ramsden estate's Almondbury properties. Here it will be recalled the estate pursued a more open policy towards mining and as a consequence the colliery numbers had increased to serve the growing industrial and domestic market. In effect the coal mining industry under the influence of a single dominant landowner was reconfiguring; collieries had to be established on the peripherary rather than within the market. This is reflected in the production figures shown below as an estate based mining industry originally created for the domestic market was increasing production to meet industrial demand and fill the gap caused by the disappearance of collieries from the central area of Huddersfield.

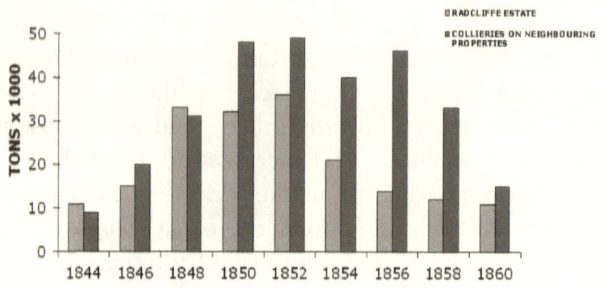

Production Levels of Collieries on the Radcliffe Estate and on Neighbouring Properties 1844-1860

In Huddersfield the effects of the Ramsden estate policy, the refusal to grant mining leases within a half mile radius of the town centre can already be seen. The collieries shown were all operating on moderately large leaseholds by the standards of the District, being more than five acres. These were primitive operations, relying in the main on increasing the number of pits to increase production. All were at the limits of the definition of 'small' collieries, being capable at that time of producing around 10,000 tons per annum. They reflect then the rapidly growing industrial and domestic demand for coal. Not all the collieries shown lay on Ramsden estate property. Those to the north were working leaseholds on Thornhill and Radcliffe land. Other collieries were present but operated at the lower end of the productive range for small collieries, usually producing less than 4,000 tons per annum. These collieries occupied land which in 1840 was still agricultural in character. They had in effect evolved from the small and primitive workings that supplied the local domestic market. At this point in time the potential for higher quality housing on the northern outskirts of the town was not yet apparent. Some of the collieries had evolved into locations capable of producing in excess of 5,000 tons per annum. Since the local domestic market was not large enough to absorb this kind of output it must be assumed that most of the production was going into the town. The linear spread of collieries to the east of the town may be explained by the need to provide coal to the industrial activities that were by now beginning to expand beyond the Holme Valley. Taken as a whole the colliery distribution pattern in Huddersfield at this time was shaped by the Ramsden estate's refusal to permit mining in the central area of the town. In effect the distribution pattern shows a coal mining industry positioning itself to work round this obstacle in order to reach local markets.

In Halifax there were two strands emerging. In the southernmost corner lay the Shibden Hall collieries and running westwards roughly along the line of the Hebble Valley were several other sites all positioned to take advantage of growing industrial demand in the valley. This of course reflects a colliery location pattern that came into being before railway penetration of the market had taken place. To the north was another line of collieries providing the link between mining activity in Halifax and Bradford. Again small settlements in this part of the District, such as Northowram and Southowram had been involved in the pre industrial working of coal. Demand in Southowram came from Halifax, whereas Northowram lay within the market area of the iron companies. It is not surprising therefore to see a denser concentration of collieries further to the north. None of these northerly locations was involved in sophisticated mining. To meet the demands of the iron companies the usual expedient of increasing the number of pits in operation was followed. The Northowram mines operated on land owned by the Savile estate which gave rise to an access problem. The modest level of mining activity, sufficient to meet domestic demand, had not been seen by the estate as financially significant. As a consequence mining was viewed as a tenant's right until rising industrial demand prompted a change in policy. However to enforce this estate was compelled to seek a legal remedy in order to overturn a practice said to date back to the reign of Henry VII[5]. Only by doing this was the estate able to grant leases to the Bowling Company.

It should be noted that market areas were not the only factor at work in shaping the colliery distribution pattern. The thickness of coal seams varied considerably and faulting could entirely deprive some locations of coal[6]. This factor also shaped decision making. For example collieries on the Radcliffe estate between Huddersfield and Halifax were working only a single seam of about two feet in thickness. This may have influenced the decision making which led to a rapid exploitation of the coalfield via short leases of under five years, using the most primitive (and wasteful) mining methods which led to considerable surface damage. Given the presence of a seam of only marginal economic value, the estate may have decided that the best course of action was to gain revenue from the leasing of mineral rights whilst the opportunity still existed, before railway penetration made the working of these reserves uneconomic.

The colliery distribution pattern in Huddersfield was to remain stable through out the 1840s, with Ramsden estate policy being the factor which maintained this. Whilst it might be assumed that the arrival of a main railway line would be the cause for change in fact it occurred due to the construction of a second route. In Almondbury to the west of Huddersfield where the Ramsden estate made no restriction on mining, the route taken by the Huddersfield-Sheffield Junction railway was to form the southern boundary of coal mining activity in the township. To the north of this line settlements such as Farnley Tyas, Newsome, Berry Brow and Lockwood had become by 1854 sites for intense mining activity, all of which was being undertaken by collieries producing less than 10,000

---

[5] Savile Estate Papers
[6] Green et al *op. cit.*

tons per annum. Beyond the railway mining was to rapidly disappear as points for the distribution of coal from outside the District were established[7].

A similar effect can be detected adjacent to the route of the Manchester to Leeds railway. The various settlements along the line, gradually being absorbed as suburbs of Huddersfield had seen the opening of new collieries as industrialisation took place. South of the route taken by the railway three collieries were working. It would appear then that the of the railway was sufficient to prompt colliery owners to establish mines beyond the points at which coal brought in from outside the District could be easily distributed.

Reference has already been made to the coal mining industry along the upper Holme valley. In 1845 there were blocks of collieries each of which served a particular settlement. The greatest number of these lay around Holmfirth[8]. Nine years later almost all of these had disappeared. The few collieries that remained were situated at points where the Holme valley was at its widest and some settlements still remote from the railway. Coal distribution facilities installed by the Lancashire and Yorkshire Railway at the small stations along this branch line were usually modest in scale; consequently a colliery lying close to a settlement might be in a good position to retain its market. An example would be the village of Honley where the greater part of the settlement lies on the north side of the valley, whilst the station lies on the south side nearly half a mile away. Here the steep northern side of the valley provided a barrier and three collieries continued working for several years after the arrival of the railway[9].

The continuum between mining in Huddersfield and Halifax crossed the properties belonging to the Thornhill

---

[7] Wray D.A., *the Geology of Huddersfield and Halifax* HMSO (1930) pp221-224

[8] Wray *Ibid.*

[9] White B., *A History of Huddersfield* (1953) p 90

and Radcliffe estates and these were experiencing different operating conditions. It has already been suggested that the Radcliffe properties were rapidly exploited to win the coal before the market disappeared. This was a marked contrast to mining taking place little more than a mile away where investment in more sophisticated mining methods was to bring only limited rewards. On the Thornhill estate two collieries were in operation and this remained the case throughout the decade. Because the Thornhill collieries had been denied access to industrial markets in the Colne Valley they were isolated and did not perform to expectations[10]. The market which existed on the north east outskirts of the town evolved into one which was mainly urban in nature and there was no incentive to increase production.

It will be apparent that small scale mining was inefficient in the sense that it the wasteful methods used could leave much coal unworked. A further problem was the amount of surface damage it caused. All collieries even small ones generate waste. Because of the primitive level of operation the ability to build waste heaps to any great height was lacking. As a consequence small tips were the usual means of disposal. This had the effect of spreading the waste over a wider area. Whilst the intrusion caused by the colliery itself was usually quite small and might not affect the land which was in close proximity, a disproportionate area was required for waste disposal. Even a colliery producing less than 5,000 tons per annum could generate a considerable amount of waste that needed room for disposal. It was a further reason for landowners to weigh their options carefully where the potential for alternative land use existed. One of the few physical reminders of the existence of the industry are the characteristic spoil heaps generally less than fifteen feet in height which may be found in villages where the industry was operating. The low height of these mounds and the

---

[10] Thornhill Papers

area of ground they covered still provides physical evidence of how much land might be needed for disposal purposes, even when only a modest amount of coal was available for extraction. It meant though that mining activity could render the surface above leasehold useless for anything else and it is unlikely that this would have been allowed by owners of anything other than low grade agricultural land or where the potential for building schemes was small[11] ; hence the Ramsden property in Almondbury was heavily exploited since there was little prospect for other types of land use. The practice of granting short five year leases, seen as beneficial to both landowner and coal master with limited funds, helped achieve this. It meant that it was relatively straightforward for the Ramsden estate to bring about the cessation of mining activity in or close to the central area once it became obvious that rising land values made the coal reserves a less attractive source of revenue and conversely in the rural parts of the property that exploitation would increase income. In this respect primitive mining methods provided flexibility. More sophisticated mining methods requiring greater capital investment would have required longer leases to enable coal masters to achieve a reasonable return. Land values within the central area had increased so much that an extractive industry was in no position to compete[12]. Royalties from mineral rights would have been far less than the income derived from urban and industrial developments. Some collieries did continue to operate within what was to become an area dominated by supplies brought in by the railway. These though were on land owned by mill owners who were taking advantage of the opportunity to work coal which lay on their properties.

In Halifax changes to the colliery distribution pattern came rather more slowly. Whilst in Huddersfield during the 1840s the movement to the periphery was pronounced

---

[11] Ramsden Estate Papers RA5

[12] Springett A.F., *The Mechanics of Urban Land Use Development 1770-1911* unpubl. Ph.D thesis Univ. of Leeds (1974)

and rapid, in Halifax the linear pattern shown in figure 1 of chapter three persisted. Whilst small collieries could be found throughout the borough those to the west of the line marking the boundary of the District were mostly those established to serve the domestic needs of isolated settlements. Bearing in mind the difficulty of moving coal over often steep topography of Halifax these collieries were the last examples of the kind of mining activity which had existed for centuries. It was only within the boundary of the District that Halifax collieries had expanded to meet industrial demand. Apart from the clustering on the Lister property which shows the location of the collieries that had been among the first to increase production to meet industrial demand, there was a line of collieries established a short distance to the north[13]. These collieries remained in operation throughout the decade despite the arrival of the railway. In contrast to Huddersfield there was no dominant landowner with control over the central area of Halifax. Rather there was a landowner, Lister, with property in close proximity whose major activity was mining. In short local coal masters, particularly Lister dominated the market where industrial expansion was taking place. This may help to explain why railway based distribution points were slower to be developed in Halifax with far fewer coal merchants operating from these sites than could be found in Huddersfield. Only in the 1860s did the number increase[14]. For the Shibden Dale collieries the range of customers, investment in more modern mining methods and greater productivity made them capable of retaining their markets. Throughout the decade Halifax collieries grew in number. Changes in the distribution pattern can be matched to the contour lines with new collieries established after the arrival of the railway locating on higher ground. The difference then between Huddersfield and Halifax was that

---

[13] Trigg W.B., 'The Halifax Coalfield' *Transactions of the Halifax Antiquarian Society* (1931-32) p56
[14] Baines *Trades Directory*

the distribution pattern did not evolve in pursuit of new markets but rather to get beyond the point where supplies brought in by the railway could be easily distributed. Competition for land use mid century was less severe in Halifax. The use of mining leaseholds of modest acreage coupled with the ability to find niche markets away from the central area allowed small scale collieries to operate and avoid competition for land. Where collieries dominated close to the central area this was chiefly on the Shibden Hall estate where mining was seen as the principal form of land exploitation.

North of Halifax the townships of Northowram and Southowram both saw an expansion in colliery numbers. There was a dividing line between the two with Southowram fitting the pattern for other Halifax collieries, establishing a mixed industrial and domestic market protected from railway competition by its elevated location. In contrast Northowram was close enough to be within the sphere of the Bradford iron companies[15]. At the time these were small isolated settlements and as a consequence mining was able to operate without coming into conflict with other forms of land use.

In Bradford Dale colliery numbers continued to grow during the 1840s. There was nothing systematic about this expansion, settlements such as Wibsey and Shelf rapidly became transformed into mining settlements. This intense clustering of sites close to the ironworks reflects the demand for coal and ironstone with a combination of deep mines and shallow pits being used to meet demand[16]. The colliery distribution pattern shows the scramble for mineral rights between two competing companies, each linked to a prominent local landowner. Smaller landowners found that their coal reserves were also in demand whether they were leasing to others or working themselves. Such was the demand that even the smallest acreages were of interest. This is reflected by the fact that

---

[15] Trigg *op.cit.* p57
[16] Cartwright J., *A History of Low Moor* Bradford (1906) p 110

in 1850 there were 70 pits sub contracted by the Low Moor company. Most of these were primitive operations working the shallower seams[17]. There was a technical upsurge in mining in Bradford Dale directly under the auspices of the iron companies' own collieries but this expertise was developed alongside mining methods that had been in use in the pre industrial era. The association of the Black Bed ironstone with the Better Bed coal seams meant that Bradford Dale was the only location in the District where mineral exploitation was the dominant factor in land use exploitation, accentuated by competition between the two companies. During the two key developmental decades, the 1840s and 1850s mining continued to be the dominant land use activity. The availability of minerals within a four mile radius of the works irrespective of acreage was the only significant factor affecting the colliery distribution pattern. The pre industrial landownership pattern had given rise to the creation of many small fields. In turn this meant for a fragmented pattern of ownership. Small landowners wishing to exploit their mineral rights found that they had little choice but to lease to the companies or work with pit takers who were contracted to supply to the Works, since the two dominant landowners were closely linked to the iron companies. Individuals with smaller properties surrounded by Beaumont or Savile lands were often compelled to lease to the companies or face problems with wayleave[18].

The railway arrived in Bradford Dale via two lines opened in 1848 and 1850 but there was no impact on the colliery distribution pattern. Indeed some Bradford coal was soon being moved out of the city. It should be appreciated that a study of the entire Bradford mining industry is beyond the scope of this study but at the height of production Bradford output was about 10% of the Yorkshire total. Only after 1849 as the shallower reserves

---

[17] Cartwright *Ibid.*
[18] Beaumont Estate Papers

began to be exhausted was coal brought in by rail to supply the ironworks.

The distribution pattern in the year 1840 marks the start of a phase in the development of the mining industry which was to continue for about fifteen years. It was the most critically important phase since it covers the years when the railway network was yet to reach all parts of the District and local supplies of coal were of great importance. A coal mining industry developed along more modern large scale lines could not have been reconciled with other forms of land use. The heavy capital investment required to sink and work a large colliery could only have been contemplated where sufficient land was available for infrastructure and waste disposal, together with coal reserves at sufficient depth to reduce the risk of surface damage. This was not the case in Huddersfield where rising land values combined with the spatial needs of industry and housing made mining a necessary though intrusive presence. By denying the central area of Huddersfield to colliery workings the Ramsden estate might have jeopardised the viability of the industry. This did not happen because small scale mining with only modest start up costs was able to insert itself more easily into any available space. As a consequence it quickly adapted to change and was able to create new market areas. The loss of the central area following the arrival of the railway did not prove damaging to the industry. As will be shown, it continued to adapt to changing market conditions throughout the remainder of its existence.

In Halifax the reverse was true. Due to the steep topography there was less available space. The dominant position of the Shibden Hall estate with mining as its priority meant that larger scale operations were possible and the local industry was able to resist the arrival of competition from the railway and retain a market share. Smaller collieries to north of the estate also benefited from this.

Two of the three major factors helping to shape the

decision making process revealed in the colliery location pattern were now apparent: the flexibility of the industry and the role of a dominant landowner. The third factor reflected in Bradford Dale came when the two dominant landowners, Savile and Beaumont effectively combined with the major industrial enterprise to make the needs of mining the land use priority.

Chapter Eight

Operating Conditions

The labour requirements of the mining industry varied throughout the District. In Huddersfield and Halifax with the long history of mining as an adjunct to agriculture it had not been seen up to the industrial revolution at least as a primary occupation. Rather it was a form of labouring closely allied to working on the land. In contrast impelled by the requirements of the iron companies, a large labour force grew up, many of whom were housed in company accommodation. Other collieries operated by pit takers or sub contractors were worked in the same way as the more primitive collieries further south. Overall the picture was of collieries operating in much the way they had always done, taking on workers who saw mining as another form of labouring, interspersed with the more sophisticated operations using a workforce for whom the term miner was a more accurate description.

There was certainly no great financial incentive to enter the trade. Even in the 1870s a miner in a Bradford colliery could be earning only £1 per week. The stability of wage rates was maintained through the 1840s and well into the 1850s. The day wage rate varied between 3s and 4s only showing a slow increase 1858 to 4s 6d[1].

Some exceptions existed. For example the Shibden Hall collieries, whilst retaining the paternalistic approach of the rural landowner were rather more generous. Here the colliery day book reveals payments of up to 4s 8d per day during the 1850s. This was of course piece work. Miners were paid by the 'load' and the overall wage bill therefore increased in line with the output. It must not be assumed

---

[1] H.M. Mines Inspectorate Report s Yorkshire District 1871- 79

that there was anything consistent in these wages. Interruptions to mining operations would correspondingly reduce the amount paid to miners. The head frame being out of action for a time at Shibden Hall colliery resulted in an average payment of 1s being paid to miners on those days[2].

One of the benefits of paying relatively good wages was that it improved the prospects of maintaining a stable workforce. There were examples of this. Shibden Dale being one and the Bradford iron companies was another. At least in those collieries operated by the companies where the day books like those at Shibden Dale show the same individuals being employed over several years[3]. This stability could also be seen where there were less opportunities for alternative employment, such in the collieries of Southowram. Sometimes an agricultural labour force might be moved into mining because that had become the primary activity on the property. Where workers were also tenants of the landowner there was even less incentive to seek more congenial employment. Elsewhere the industry had a transitory workforce which reflected the short working lives of collieries working small leaseholds. Such collieries were not intended to remain in operation for more than a few years and closure might mean that the workforce had to seek alternative employment unless the coal master could find a new leasehold nearby in which case the miners would commence a new sinking.

There is no evidence that mining labour became organized. Records of the various mining unions show no branches in Huddersfield, Halifax, or surprisingly even in Bradford Dale where it might be assumed that a workforce running into hundreds would see the attraction of combining. Elsewhere it is easy to appreciate that a wide scattering of small collieries with a transient workforce would have made it difficult to organise, even if the will to

---

[2] Shibden Hall Papers CMA4
[3] *Ibid.*

do so had existed.

The turnover of the labour force recorded in colliery daybooks can be attributed in part to the conditions underground and also the uncertainty of the work. As noted earlier, underground conditions often interfered with production and where a single coal face was being worked this would mean that all production ceased. At Lane colliery in Huddersfield in May 1859 the miners were laid off following an inrush of water from abandoned workings. Pumps were brought in an attempt to drain the mine. Two weeks later the mine commenced working again but of the fifteen men employed before temporary closure, only five of the originals remained[4]. There was a similar situation at New Bank Colliery in Halifax where in 1862 the seam vanished. Production ceased and the dozen miners were immediately laid off. Mining resumed the following month when the coal master had negotiated the sinking of a new pit with the landowner. However only six miners were needed to actually do this work. This was one of the principal shortcomings of primitive mining; being erratic in nature meant that the retention of a stable workforce was a rarity. Drawing upon surviving colliery day books from Huddersfield the turnover in workers over three years in the 1860s shows that only 20% of the original workforce was still employed at the end of this period[5]. An examination of census data for the years 1841 and 1851 in selected Huddersfield and Halifax wards supports this. There was no concentration of miners even in enumeration districts where mining was taking place in close proximity. The only significant mining communities were those close to collieries operated by the iron companies. Here there were sufficient sizeable collieries of the type to have lengthy working lives operating on extensive leaseholds prompting the need to recruit, house and retain a sizeable workforce for whom mining was a settled form of employment rather than an occasional

---

[4] Huddersfield Tolson Museum, colliery records
[5] Huddersfield Tolson Museum, colliery records

occupation. The Shibden Dale collieries came close to this category but can also be seen as an example of a landed estate diversifying away from agricultural activities, where some land workers redeployed into mining, augmented by others from the town as demand for labour increased. With the exception of the ironworks, no colliery companies of the kind familiar in other parts of Yorkshire were established. Essentially geological factors kept such companies away. Whilst there was a market for the coal, investment in deep mining would not have been worthwhile, consequently the industry remained largely in the hands of small scale coal masters seeking niche opportunities to exploit modest acreages of mineral rights.

The economic value of the coal coupled with the depth at which it was found often determined the level of investment. The iron companies' leaseholds in the east of Bradford Dale were working coal at depths of 240 to 250 feet, which precluded anything but deep mining techniques. In such locations steel headframes and steam powered winding machinery and ventilation were the norm. These were the locations where the metallurgical better bed coal was to be found. In the west of the Dale coal was to be found and shallower depths, between 150 and 220 feet, again too deep for anything but advanced mining techniques to be used.

The records of Listerwick Colliery on the Shibden Hall estate provide a detailed picture of operating conditions and costs during the 1850s. This was a colliery going through a transitional phase, expanding and modernizing into a large producer. The colliery accounts for 1843 reveal that it was still operating in the traditional manner of an estate colliery making sales of between two and 160 'loads' of coal to householders and coal merchants distributing through the vicinity. Some miners were producing as much as 22 loads per day, others as little as two or three. Judging by the turnover of names in the colliery day book these discrepancies may be explained by the recruitment of inexperienced miners who were yet to learn techniques of underground working[6].

During 1845 the mine worked for 50 weeks of the year, indicating that demand was good and there were no interruptions to production caused by geological or other factors. This high level of production was to continue on into the early 1850s. Some of the sundry expenses recorded give an indication of underground operating conditions. Pump leathers, suggesting a shallow mine using crude methods of drainage, rails and sleepers indicating that the colliery was spending money on underground transportation. This also suggests that the workings were now becoming more extensive. In 1846 the annual income of the colliery was £234 and thereafter it rose significantly. A half yearly profit in excess of £1,000 was recorded each year up to 1851. Whilst the colliery was now focusing on industrial customers an inferior seam, described as 'slack coal' was still being mined for domestic sale. The average wage for a miner in the early 1850s was 17s 6d per week with an overseer receiving 3s 6d per day. Payments to an 'engine tenter' reveal the introduction of steam power. By the 1860s the annual wage bill had risen to £800 per annum, marking the transition to a sophisticated mining operation. Profits also increased; by 1862 the mine was earning £3,889 for the estate[7].

Running a colliery with mechanized winding machinery meant that some of the coal used was directed to the engine house. In 1866 colliery accounts show that consumption was 46 loads per day. The local measure of a load on the estate being 40 loads equaled four tons. Again this indicates a lengthy working day with the winding engine being well used. The mine manger noted in correspondence that tree cover on the estate was being rapidly denuded to provide timber for the collieries, a further indication of how the mining activities had by far become the dominant form of economic activity on the

---

[6] Lister Estate Papers CMA5
[7] Lister Estate Papers CMA5

estate. This provides an example of how the needs of an extractive industry were dominating land use in this part of Halifax. The strength of the Shibden hall collieries in the local market even at a point where the railways were well established is indicated by this.

A further benefit came from the presence of fireclay. Again in 1866 this was being mined at an average rate of 260 loads per week. Each load being valued at 5d. There were though breaks in production and the impression is that this secondary mineral was worked intermittently, with sales being made to local brick makers.

By the 1880s production at the Shibden Hall collieries had begun to decline. This can be measured not only in production but also miner's earnings. By that point very few were earning more than £1 per week. Operating costs had also declined and ranged between £15 and £30 per week. Records show that in 1881 a total of 808 shifts yielded only 1,079 tons and significantly the estate was seeking to lease mineral rights for the first time. Ten acres of 'soft bed' coal were obtained from a neighbouring landowner at £52 10s per acre in 1885[8]. This was a significantly higher figure than the District average at that time and may reflect the fact that the lessor was in a strong negotiating position, presumably aware of the decline in reserves on the estate and the strategic value of mineral reserves lying in close proximity to the property.

By 1888 the first indications of the Shibden Hall collieries going into the red are apparent. The collieries also experienced an increase in the number of accidents. The sum of £50 appears to have been the going rate for the death of a miner in the District at this time and the estate made two such payments during the 1880s. Space was also becoming a problem and 'small rents' were being paid to neighbouring landowners to dump waste on their property. The collieries were proving less able to supply industrial customers and provide fuel for the ovens at the brickworks

---

[8] Lister Estate Papers CMA4

that had been established on the estate to make use of the fireclay which was being mined. The estate was then in a position where demand for coal remained high and the price locally was good. In 1858 the Hard Bed coal had been fetching 7s per ton. By 1888 it had risen to 11s. This figure was marginally less than the cost of supplies brought by the railway[9]. However with declining output and high running costs the lack of reserves on the estate and only a few leaseholds of a modest acreage available nearby, meant that the development costs of extending the workings onto neighbouring properties would reduce income until these new faces came into operation. Further the cost and effort involved was unlikely to be repaid. Arguably the ten acre leasehold obtained by the estate on a neighbouring property might have been worked more economically by using primitive mining techniques with the site remaining unconnected to the Shibden Hall colliery.

The collieries belonging to or operating on behalf of the iron companies reflect a mixture of the primitive and the relatively sophisticated. For example whilst steam power was introduced at an early stage, dating back to the installation of a Savery engine in the 18th century and subsequently the use of Bolton and Watt engines[10]. Water power was also in use. A Bowling colliery was able to raise weights of 160lb by this method. Account books also show the usual recourse to local tradesmen for repairs to such items as tools and corves, generally an indicator of a primitive colliery too small to have its own workshop. The transition from a primitive colliery using a series of pits to a more sophisticated operation was not always the best decision. In some cases it might be described as evolutionary; the demand for coal was high, prices were good and rather than hire more miners to sink additional pits, the workings might be extended. However in a

---

[9] Brigg J. The Industrial Geology of Bradford(1908) p92
[10] 'Benjamin Gott and the Industrial Revolution in Yorkshire', *Economic History Review* (1931) p34

shallow coalfield wet conditions were often a problem. Lacking steam power and the ability to drain the workings mechanically meant that soughs had to be dug. In effect the saving made by not excavating vertically might be lost by having to dig horizontally.

Both of the Bradford iron companies relied heavily on 'pit takers' independent coal masters used to augment production. The advantages of this approach were twofold. Firstly a pit taker was ideally suited to work small leaseholds often of five acres or less. Such a small acreage would not have repaid investment in more sophisticated mining methods and therefore the coal would probably have remained untouched but for the ability of a pit taker to utilise primitive mining methods. Secondly the workforce was the responsibility of the pit taker. Inevitably those miners hired to work in such collieries laboured under more difficult conditions than those directly employed by the iron companies. For the lessor the pit taker system provided advantages. The demand for coal and ironstone was to produce considerable revenue for Savile and Beaumont the two largest landowners involved with the companies and this was enhanced by the ability of the pit taker to exploit small parcels of land.

The length of leases varied and there were three classes: fixed sum per ton raised, sum per acre worked and a sliding scale varying with the selling price of the coal raised. In Yorkshire it was based on acreage and sometimes seam thickness. Rates of royalties at times when trade was good meant a lease taken out during this period may have a higher rate of royalties. The quality of coal was also an influence. Average rate of royalty on coal per ton in Yorkshire was 6d [11]. Where 'short workings' occurred it meant that the amount of coal raised was less than the fixed rent. The difference could be recovered by overworking in the following year. The estimated charge

---

[11] Report of the Royal Commission on Mining Royalties ( ) pp 1,893, 1,894

for royalties in Yorkshire in 1889 based on 21,976,027 tons raised £3,549,401 which indicates what a huge source of income it had become for landowners. Mine owners often argued with their workers that royalties govern wages but in fact when first taking out a lease the lessee must take into account the wages and development costs he must pay so wages in fact may be said to govern royalties[12].

On the Shibden Hall estate Listerwick Colliery maintained a total half yearly income of £1,000 or more throughout the 1870s indicating that this was the most productive colliery on the estate and had not yet begun to experience a decline in reserves. However indications that the colliery was experiencing pressure on its market share became evident. Sales to major consumers such as the local gas company continued and also to the well known Halifax firm of Crossley and Leeming but the practice of selling coal at the pithead for domestic consumption resumed, suggesting a surplus of coal requiring a market. Maintenance tended to be carried out by jobbing tradesmen and clearly there was no reluctance to put individuals to work in what must initially have been an unfamiliar environment.

Wage rates were also be affected by the type of coal being mined. Where there was a demand for good quality engine coal that fetched the highest price the miners benefited correspondingly. On occasions they might be working seams of the poor quality 'slack coal' as it was known locally. This was the coal sold for domestic use. Other work such as shaft sinking also affected miners' earnings as this paid less than actually producing coal. Part of the problem was the sporadic nature of purchases. Through the 1870s the Shibden Hall colliery day books reveal a growing number of sporadic purchases in modest amounts from a variety of customers such as engineers and

---

[12] Aldcroft D.H. The Entrepeneur and the British Economy 1870-1974, *Economic History Review*, 2nd Series (1934)

brass founders. In effect the collieries were returning to the type of operation from which they had emerged two decades earlier. This illustrates how in Halifax the local mining industry was of diminishing importance. The supply of coal via a contract was insufficient to sustain the operation and as a consequence there was a growing dependence on one off purchases.

As the collieries had been developed and modernized so the estate began to integrate other activities, for example the manufacture of sulphuric acid from pyrites. This appears to have been done to order since entries in colliery day books referring to coal being used for this purpose are sporadic. Another option was the mining of fireclay but this was a low value mineral in contrast to coal being worth only 5d per load. However when demand was high miners could earn relatively good wages and of course the clay was more easily worked than coal[13]. These examples of vertical integration illustrate how the estate attempted to be more than simply a supplier of coal. In undertaking these activities the estate was able to establish a niche in the market for forty years, even in the face of railway competition. However such activities were probably undertaken sporadically and on too small a scale to be a viable complimentary activity.

Other than those operated by the iron companies the Shibden Hall collieries provide an isolated example of colliery operations beginning at a primitive level based on their agricultural origins and evolving towards the limits of technical expertise to be found in the District. The juxtaposition of the primitive and advanced never quite disappeared owing to the need as reserves were depleted to find opportunities for leaseholds nearby. Here we reach the central problem of mining in the District, that where the expense of developing new workings was too great to justify linking these to the original operation it proved wiser to resort to primitive methods of mining, which were

---

[13] Shibden Hall Papers CMA4

of course wasteful and a far less efficient means of working the coal. In short the price paid for an acre of coal to be worked was unlikely to provide an actual acre raised. However as competition grew stronger the estate found it necessity to resort to this method in order to maintain production and meet the needs of its customers. The paradox was that as it sought to do so, the customer base was declining due to competition from the railways. As noted earlier this was evidenced by a resumption of pit head sales for domestic consumption and the erratic nature of sales to industrial customers. The Shibden Hall collieries had by the 1870s ceased to be at the forefront of mining operations in Halifax and the use of local tradesmen to supply equipment and services to the collieries marked a reversion to the practice at more primitive collieries in order to reduce operating costs.

The remaining exceptions to this were the iron companies particularly after 1870 where technical expertise was to be found. The companies were something of a centre of excellence, willing to loan their engineers and surveyors to other colliery owners. There was of course an element of self interest in this since the companies would probably have had an interest in purchasing the coal mined, particularly if it was found in association with ironstone. Whilst underground conditions may have been marginally better at the iron company collieries there were limits to how far investment in new techniques was to go. In common with much of the British coal mining industry, post 1860 there was no use of mechanical coal cutting machinery[14]. Arguably there was a drag on how far the companies needed to go in upgrading their collieries when production could be increased via sub contracting. The willingness of the companies to buy coal from any source is illustrated by the fact that even in the 1880s, close to the Low Moor works in Wibsey it was still possible to find coal being extracted from day holes[15].

---

[14] Jevons W.S. 'The Coal Question '(1906 edit.) p211

[15] Briggs I. op.cit. p72

This disorganised approach extended to surface operations where both companies ran tramways of differing gauges. This was probably prompted by the need to extend out quickly to small leaseholds where the working life of a colliery was not expected to last more than a few years. The fact was that throughout the century and despite the eventual spread of the railway network to all parts of the District, the small colliery continued to be of value. One important reason for this was than in larger collieries producing more than 10,000 tons of coal per annum it was inevitable that a proportion of the workforce would be ancillary to the actual business of coal getting. In the small colliery almost all of the workforce would be employed in working the coal. '*A small owner working shallow seams whose capital resources are limited is able to compete with large, deep, well equipped mines. The ability to do so is no test of efficiency. Productivity (output per man employed per unit of time) is an imperfect measure of efficiency. The real test is the extent to which actual performance measures up to the potential performance under particular physical circumstances*'[16]. Shallow pits may then have been an inefficient means of working but there was a balance to be struck and certainly before 1860 in Bradford Dale as elsewhere in the District, this was tilted in favour of the primitive colliery. Investment in deeper mining was a feature that emerged later in the century. Prior to 1860 with small acreages available for leasehold it was easier and cheaper to employ a pit taker and by this means production could be increased quickly. Where pit takers were used the work was undertaken on the basis of so much being paid per dozen corves. During the early years of the nineteenth century the iron companies did improve efficiency at these primitive workings by the use of tramways. The contents of the corves were tipped into wagons which were then drawn away by means of a stationary engine. Such sites

---

[16] Taylor A.J.. '*The Coal Industry*' in 'The Development of British Industry & Foreign Competition 1875-1914 ' (1968)

were temporary but the method proved useful for linking pits working small acreages on scattered leaseholds and of course had the benefit of extending the market area of such collieries.

A miner working the Better Bed seam which was of greatest value to the iron companies could in 1845 earn 2s 10d for producing 54 hundredweight of coal. Bearing in mind the thinness of the seam this was not a generous rate. However they fared better than those mining iron ore. The working day could last between 12 and 16 hours. Generally those directly employed by the iron companies were better off than their colleagues working elsewhere. As with the Shibden Dale collieries the iron companies were prepared to pay compensation as a result of accidents and accommodation was also available for some. The worst collieries were those operated by what might be termed speculators. In short where mining was a subsidiary activity. There are accounts of weavers, joiners and grocers leasing a few acres of poor quality agricultural land and then covering these with day holes in which a few miners would labour in poor and often dangerous conditions.

Where seams were relatively thin it became customary to let the coal reserves by the acre. Agreements would include a fixed rent, often the major operating cost in the more primitive collieries, together with a royalty. When output fell meaning that the royalty was less than the fixed rent this was known as 'shorts' and the lessee was sometimes able to deduct it from the sums payable for output in excess of the minimum in subsequent years. At one leasehold in North Bierley the Low Moor company was required to pay 50s per annum for every acre of land used and taken up for the working of coal until 'the land should have been made fit for ploughing'. From this it may be assumed that primitive pits were used and the land 'restored' afterwards by dumping colliery waste down the shafts.

To create a colliery capable of producing between two

and 10,000 tons of coal per annum was not a particularly expensive undertaking. Leaseholds were at the rate of £40-45 per acre; a figure which remained remarkably stable throughout much of the century. This constituted the major start up cost for a small coal master. The equipment costs could also be modest.

The rapid expansion in coal mining during the 1840s and 1850s was accomplished chiefly by means of expanding the number of sites being worked. In addition the retention of primitive methods was not simply evidence of clinging to an outmoded pre industrial approach. Rather it was crude though effective means of working coal in locations which because of their small area could not otherwise have been exploited. In turn this created greater scope for industry to expand into districts where before the arrival of the railway fuel supplies would have been difficult or expensive to obtain. In Bradford Dale demand came overwhelmingly from a single source and here exploitation occurred more quickly because the iron companies were prepared to create a transport network that extended their reach. Without this dominant customer it is likely that the thin seams of Bradford Dale would have been much less attractive and as a consequence collieries in the area would have felt the effects of railway competition. Instead they remained in operation until reserves were exhausted. Rapid exploitation came with a price. Apart from inefficient working the chaos caused by different methods of mining was to prove expensive to the iron companies. This was revealed in the 1880s when after several decades of rapid expansion a full survey of the Bowling mining operations was carried out. The ensuing report vividly illustrates how the need to extract coal and ironstone had overridden other considerations. There were several examples of where mining had encroached on neighbouring properties and as a consequence the company was forced to pay compensation. The varying gauges used on the tramway network caused difficulties when transporting and unloading coal. However by that stage both the Bowling and Low Moor operations had reached the limits of their expansion and whilst some of the less productive collieries were closed the survey was arguably conducted too late to bring about real improvements in efficiency.

Colliery operating conditions in the District were shaped by market requirements and indirectly formed a further variable in the decision making process when decisions were taken affecting land use. The modest start up costs, combined with the ability to render the working of small acreages of mineral rights made the primitive colliery well suited to an annual production rate of less than 10,000 tons per annum. An unskilled workforce was capable of being employed in these collieries, partly because legal protection was minimal or not well enforced. The transition from agricultural worker to miner was easily made for example on the Shibden Hall estate where this provided the original workforce. Nor did coal masters have to concern themselves with a workforce that was unionised. Further, as has been shown, there was little in terms of productivity to separate the miners employed in primitive collieries from those in the more sophisticated operations. Ultimately for the landowner not directly involved in mining, the type of operation could be varied to suit the extent of the mineral rights. The dominant industrial activity tended to play a large part in shaping the decision making process for the landowner. A greenfield site retained for industrial development might still contain a parcel of land where coal might be worked using primitive conditions, suiting both landowner and mill owner. In the north of the District where industry in the shape of the iron companies dominated the decision making process, the economic reality for landowners such as Beaumont and Savile meant cooperation and mining as the primary form of land use. Operating conditions were then shaped by the size of the leasehold.

Chapter Nine

Production and the Role of the Industry After 1855

It will now be apparent that for landowners in the District a complex array of factors might have to be taken into consideration when mining as a form of land use was being contemplated. An additional complication was that these factors could change over time. The primary role of the coal mining industry prior to 1855 was to assist industrial take off and the expansion of the urban area ahead of the arrival of the railways. This then was the era where the decision making process was at its most complex. At this penultimate stage of the analysis they may be summarised as follows:

The first option was to *Do Nothing* in effect continue to be an agricultural landowner dependant upon farm rents for an income: theoretically possible though it should be appreciated that the prospects of earning a reasonable income this way were much reduced by the middle of the nineteenth century as imports of cheap food increased. The encroachment of other types of land use into what had become a densely populated area made an agricultural estate a less attractive option. Added to which other forms of resource exploitation were very closely allied to agriculture meaning that mineral extraction was already an integral part of estate revenue. The only real advantage of this option was to leave the land intact, avoiding some of the more destructive aspects of mineral working perhaps in the hope that more attractive opportunities might arise in the future. This option did not sit well with the prevailing sentiments of the time and is hardly surprising. The owners of estates within the District were the inheritors of land which had for centuries yielded only modest incomes[1]. All of the principal landowners with the exception of the Lister and Thornhill families had more attractive properties and the principal mansion house located elsewhere. The Beaumont family for example owned Bretton Hall which lay in the midst of parkland near Wakefield. Sir John Ramsden the fifth baronet preferred to reside overseas and the Ramsden baronets were the owners of Rudding Hall near Harrogate. None of these owners would have felt directly threatened by industrial encroachment and therefore the desire to preserve property intact for aesthetic reasons did not arise. The smaller landowners with declining agricultural incomes had an even greater incentive to select from one of the alternatives. However they were not always in a position to influence the shape of developments. Location could be an important consideration. This is illustrated by the Thornhill family who wrongly anticipated industrial expansion towards their property on the north eastern outskirts of Huddersfield and invested in sophisticated colliery ventures that were to find no market opportunities to match their intended productive capability.

The second option was *Develop for Building*. This was an approach that could have a significant effect on the prospects for resource exploitation, since the two were rarely compatible. At first glance in an area where agricultural land was of low value and mainly used as pasture, the chance to utilise what lay beneath the surface to generate revenue was attractive. The dominant landowner could have a strategic role to play and whilst this was of course done out of self interest its effect on the coal industry could be significant. The Ramsden policy of driving mining out of the central area of the town had the effect of accelerating railway penetration of that part of the market. Collieries operating on the periphery could not compete with the railway which brought points for distribution right into the heart of a market.

---

[1] Gregory D., 'The Process of Industrial Change 1730-1900' in *An Historical Geography of England and Wales* (1978)

Given the steep topography to be found in many parts of the District, gradient could also be an important factor. For example the expansion of Halifax along the floor of the Hebble Valley meant that reserving land for building purposes was the preferred option. Mining was relegated to the less favourable sites above the town. This had been a pre industrial practice originally to meet the needs of once isolated communities and which expanded in the industrial age.

The final option was to *Prioritise Mining*. For example such was the wealth of mineral resources on Beaumont and Savile properties in proximity to the iron companies that no other option would have delivered anything like the same income. By adopting this approach the landowner was in effect accepting the damage that was an inevitable accompaniment and would reduce the prospects for any other form of land use except peripherally.

Either of the first two options could present a problem for the coal master seeking leaseholds close to a market. The response to this was to *Relocate*. For the landowner this was both a risk but potentially an advantage. An extractive industry is usually regarded as static, because of the need to install equipment and infrastructure. Given the primitive nature of much of the mining industry in the District plus the modest start up costs, this could work to the landowner's advantage. For example on the Radcliffe properties to the north of Huddersfield; perhaps a reason behind the granting of short duration leases. In theory the lease would bind a coal master to the property for a term of years. In practice though, once a market had disappeared there was little the landowner could do in the face of economic reality. Once denied a market by the arrival of the railway or other significant change in operating conditions, then the limited market area of the typical small colliery made it impossible to seek other outlets for sales. The small colliery paying a modest annual rent of less than £50 per annum for each acre and with little in the way of equipment or infrastructure was ideally placed to cease operations in one location and move elsewhere. This contradicts the view that investment in a colliery was both literally and figuratively sunk costs and if the market disappeared then so did the colliery. There is a clear correlation between the closure of collieries whose owners are known and the opening of new workings.

The small coal masters were transplanting collieries as market penetration by the railway companies took place. Primitive collieries were ideally suited to the concept of mobility and shallow reserves meant that the break in continuity and therefore income could be kept to a minimum, perhaps for only a few weeks before coal was being produced once more. In some cases coal masters were able to accelerate the process even further by reopening an abandoned colliery and restarting production. A site that had been abandoned might still have been viable in terms of reserves, since there were other reasons why production might have ceased, for example a lack of means to drain the workings. However if no alternative opportunities existed within the limited market area and transportation proved difficult then relocation was the only alternative. This kind of entrepreneurial activity could only occur because certain conditions existed: low grade agricultural land, a wool textiles industry that was rapidly embracing steam power and seeking new locations and topographical difficulties that affected the expansion of the railway network. This contradicts the early start thesis; that a pioneering industry will ultimately suffer from the disadvantages of an increasingly antiquated capital stock. Ironically had the coal reserves of Huddersfield and Halifax been of a better quality therefore attracting greater investment, this would have reduced the ability of the industry to respond to new market opportunities in harder to reach locations. Lacking the need to make significant investment in equipment and infrastructure the effects of an 'early start' were not apparent. What Huddersfield and Halifax required was the kind of mining that had been developed on the landed estates during the pre industrial era. 'Small outdated inefficient pits with inadequate capital resources'[2] may have been a problem elsewhere. In Huddersfield and Halifax they were an asset and the reason why the coal mining industry of the District survived as long as it did.

What occurred from the 1850s onwards was an industry that evolved and adjusted to a market which had changed due to the availability of coal brought in by railway. It was no longer vital to the economic development of the District. The 'mobile' colliery was now appearing to ensure its own survival as market opportunities continued to shrink. This option was not always available for example in the narrower Pennine valleys such as the Holme. Here there were no markets to move out to since elasticity was absent. Instead the collieries perished, sometimes in the space of a few weeks after the arrival of the railway. A two foot seam worked via a horse gin was cheap but such primitive methods of production could not compete with the reliability of railway supplies when there were no now opportunities beyond the reach of the railway..

Elsewhere as the towns expanded with outlying settlements becoming urbanized, so the coal industry could move with this inserting itself into the smallest of leaseholds, working one or two acres for example that were unlikely to be used for building. In the case of Huddersfield this persisted from the 1840s through to the 1860s.

To a lesser extent the same phenomenon can be observed in Halifax though here the attrition rate was much greater owing to the fact that railway penetration was easier and much more of the Halifax market could be reached from the main railway line which ran down the valley. In short there was less need for branch lines because industry was more concentrated. Therefore we see

---

[2] Rostow W.W. 'The Process of Economic Growth' (1960) p142

the relocation of collieries during the years 1848 to 1853 after the arrival of the main railway line. This was accomplished by moving to sites well above the town, higher up the Hebble Valley[3].

This phenomenon of colliery relocation has not previously been explored. How was it accomplished? With a primitive colliery the process of investment as already noted involved the major cost of leasehold. There might be a loss taken on this if an operation was abandoned early but essentially landowners were prepared for this hence the common practice of the short five year lease. This reflected economic reality and the fact that the coal reserves were not usually substantial. Landowners were gearing the provision of a lease to a timeframe which stood a better chance of attracting a small coal master. With no large colliery companies interested in working the reserves and the often fragmented pattern of landholding meaning that often only very small acreages were available the time frame and area became linked. This practice was also adopted by the larger landowners such as the Ramsden estate. In short early termination of a lease was unlikely to leaver a landowner seriously out of pocket. It was unlikely that a landowner would be ignorant of market realities. A small coal master was only capable of selling and delivering within a restricted market area. Were the market to disappear then the coal master was in no position to extend his market from the existing site. If there was no opportunity to move into an adjacent market then the only alternative to this was the method adopted by the Radcliffe estate; that of very short mining leases and unrestricted mining which did a great deal of surface damage and left much of the coal unworked. Essentially it was better to do this and hope that market forces remained benign until most if not all of the coal was worked out.

The closure of a colliery could be accomplished with relative ease. There was little in the way of equipment that

---

[3] Trigg W.B., ' The Halifax Coalfield' part v Transactions of the Halifax Antiquarian Society (1931) p62

had been installed in the typical small mine and therefore salvage work could be undertaken without difficulty. Similarly the surface infrastructure was also easily dismantled. Some records point to the need for hauliers to take away timber head frames that could be reused. Unsurprisingly landowners usually insisted on shafts being sealed. This requirement often led to an exchange of correspondence. Shaft filling even in a shallow coalfield would take time and manpower and coal masters were often reluctant to deploy their workforce on an operation that was not earning them any money[4]. Otherwise materials and equipment were easily removed for recycling. Although standard mining leases called for the making good of surface damage there seems to have been little attempt to enforce this, hence the physical evidence that remains to this day of grassed over spoil heaps. Essentially landowners did not force the issue on property that was unlikely to be used for other purposes since it would probably not have been given over to mining in the first place.

Once a coal seam was reached then unless a return on the investment was made quickly the colliery owner preferred to cut his losses. This illustrates how precarious the operation could be in a marginal coalfield. Nevertheless the industry was able to continue working on the fringes of the urban areas and in doing so provided support for the expansion of the town. A colliery that could insert itself into a small niche of land would be in a position to provide fuel as a service to the local domestic market if industry was not being served.

In Halifax the situation was somewhat different. Again geography has a part to play. It has already been noted that in terms of industrial and urban development the town was more compact because it was contained in a steep sided valley.

Some parts of the Halifax coal industry had operated on

---

[4] Ramsden Estate papers RA5

the higher parts of the surrounding fells providing fuel to isolated communities. Because fewer branch lines were needed In Halifax, those parts of the industry that were operating closer to the town centre were largely wiped out, with the exception of the Shibden Hall collieries. The industry that remained operated along the valley contours, taking advantage of the steep gradients to retain a market share at a point where there were obviously some advantages to be gained by being in a locality where that the coal did not have to be moved far. The pattern which emerges from the 1860s to the 1880s is of collieries locating and relocating in a linear fashion along the upper settled limits of the valley contours; in effect using gradient to remain as remote as possible from railway competition[5]. Sometimes there would be an overlap with the coal master keeping a colliery in production whilst at the same time using some of his workforce to develop a new pit. If the coal at the new site was found to be satisfactory then old workings would be closed and the remaining workforce moved to the new site. Landowners lacking the necessary technical knowledge might have to accept the coal master's view that this was necessary because the leasehold was failing to live up to expectations.

These niche markets could be unreliable. Since few small collieries could meet the entire needs of a mill engine house, purchases tended to be erratic. Apart from the Shibden Dale collieries the quality of Halifax coal was not up to 'best engine' standards of the type brought in by the railway. On these marginal sites along the fringes, the value of the local colliery was not then as in Huddersfield in assisting the wool textiles industry to expand. Though there were the myriad small factories, foundries, chemical works and the like to be found in the town and these were customers for the coal produced by small collieries. It was the domestic market and pit head sales which mattered to

---

[5] Trigg W.B ' The Halifax Coalfield' parts I & II *Transactions of the Halifax Antiquarian Society* (1930) pp44-45

these collieries. Cheap coal and a market area of perhaps a few streets were a useful combination. In general the house coal sold was of a poor quality but the price would reflected this. As in Huddersfield relocation to a new site in the vicinity when the leasehold was exhausted would retain the market. The poor quality agricultural land and steep gradients were a useful combination. This was a characteristic more of the Halifax than the Huddersfield coalfield[6].

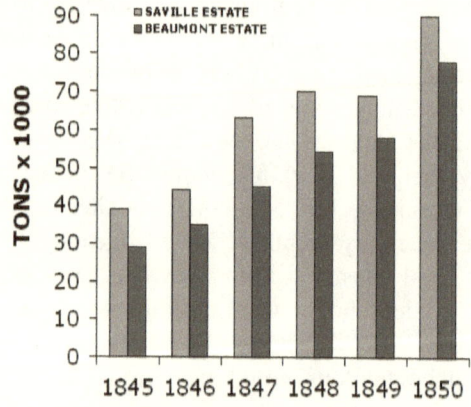

Production Levels of Coal on the Saville and Beaumont Estates 1845-1850

---

6 Trigg W.B. 'The Halifax Coalfield' parts I & II *Transactions of the Halifax Antiquarian Society* (1930) p62

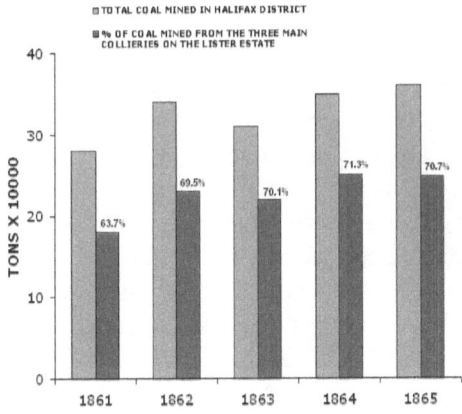

The Production of Coal on the Lister Estate near Halifax 1860-1865, compared with total output in the district.

It was this mobility or elasticity that enabled a mining industry selling what was often an inferior product mined from thin seams which enabled it not only to continue but also to increase output. The pre industrial tradition of small scale mining which aided the rise of industrialisation also ensured the survival of the industry whilst sufficient reserves remained. A form of mining traditionally seen as being of little significance was then still serving its local market some twenty years after the point where that market was still largely dependant upon local supplies. This could only have been achieved by flexibility and low start up costs. Far from being driven out of business by supplies brought in by railway the industry was able to continue working in some strength up to the 1880s and beyond this point it was geology rather than competition that put it out of business.

The production figures shown above reflect the presence in 1840 of the two significant producers in Halifax and Bradford. The demands of the iron companies and the transition of the Shibden hall collieries coupled with the lead it took in prompting the expansion of the mining industry along the floor of the Hebble Valley contrasts quite markedly with Huddersfield where the policy of the Ramsden estate in keeping mining out of the central area shows clearly how the reserve land for building option retarded productivity at a time when Huddersfield industry and population was expanding at much the same rate as Halifax. In the case of Huddersfield there was of course greater space for expansion since the town was not as constrained physically as Halifax. It shows clearly though that in the key decade the 1840s when the District was largely an Isolated Market and heavily dependant upon local supplies of coal that production in Huddersfield was still lagging behind Halifax. Had this been a coal industry which carried significant start up costs then it is difficult to see how demand could have been met locally. However the mobility of the industry allowed it to remain flexible and continue to locate as market conditions altered.

This pattern continued through the 1850s as the locational advantages of local coal supplies in Huddersfield and Halifax continued to decline. Production maintained a steady rise suggesting that demand remained high and that the as yet incomplete railway network in the District was not allowing the market to be dominated by supplies brought from further afield. A decade later and the continuing rise of production in Bradford is evident as the iron companies continued to acquire leaseholds in Bradford Dale and used a range of mining methods to obtain the coal they needed. Meanwhile both Huddersfield and Halifax were reaching the limits of their productivity. In Halifax this marks the point where the Shibden Dale collieries reached their highest point of production. The benefits of primitive mining techniques and the flexibility allowed by mobile collieries were now reaching the limits of their usefulness and the realities of geology was beginning to be asserted as fewer leaseholds became available. No landowner had been willing to leave the land idle and by the 1860s the evolutionary pattern of both Huddersfield and Halifax was established. Mining was an option of declining value and by the next decade this was to be reflected in the production figures of the 1870s. For the first time production in Halifax fell below that of Huddersfield and this reflects the efforts of the Lister estate at that time to obtain new leases on nearby sites. Huddersfield it seemed still had the capacity in colliery numbers to maintain its production figures. By this time only production Bradford was continuing to rise. Again geology was playing a part in this. The shallowness of the coalfield and the absence of deeper lying reserves meant that exhaustion was beginning to occur. However the choices made by landowners in the 1840s and 1850s were also having an impact. Whether it was to refuse to lease mineral rights, reserving the land for other purposes or rapid exploitation ahead of an expanding railway market, the reduction in accessible sites was a cause of the decline in production rather than just geology. The figures shown above reveal that the local industry was still finding a market. It was choices made twenty or thirty years earlier that were now having an impact. Only in Bradford Dale where mining was the dominant enterprise did production continue to rise.

The final phase where the industry was still playing a significant role came in the 1880s. The huge exploitative effort in Bradford Dale had now reached its limit. The iron companies were beginning to bring in coal from elsewhere and this decline was having an impact on overall Bradford production. Similarly the Shibden Dale collieries (though some mining was to continue on the estate until as late as 1910) were no longer significant[7]. It was not then the simple imperatives of markets, geology and the railways which explain the production figures but rather the decisions made by landowners in the early developmental stages.

---

[7] Trigg W.B. *Ibid.*

Chapter Ten

Strategic Choices

The discussion so far has identified the various factors affecting the decision making process when the principal landowners in the District considered what use to make of their properties. It may be recalled that the coal mining industry went through two phases, the first being the 'Isolated Market' when the advantages of a local coal supply were at their greatest. This was the point when these factors were at their most important and complex. Operating conditions did not remain static. Indeed so fluid were the factors affecting the decision making process that a landowner might be required to act quickly in order to secure the greatest advantage. The second phase came after the arrival of the railway. Such was the pace of railway expansion that the value of mineral rights might be eroded in a matter of weeks, putting the landowner under time pressure and as has been noted, railway expansion did not always follow commercial logic. Testing decision making against the precepts of Game Theory is then an attempt to unravel a complex series of variables to help explain how the mining industry was accommodated, particularly since its ability to operate during the first phase was essential. The final step before applying the test of Game Theory is to briefly review the colliery distribution pattern during the two developmental phases.

By 1840 in the south of the District the effects of the Ramsden estate policy was to push mining to the periphery, with both small (producing under 10,000 tons per annum) and larger collieries operating on the margins of the central area.

Further north in Halifax were the emerging

concentrations on the eastern side of the town, for example on the Lister estate and in the settlements of Northowram and Southowram. In the case of the latter the expansion in colliery numbers was due to rising demand in Halifax. Northowram was within reach of the iron companies and expansion there was prompted by the growing demand from that source. Throughout the District the industry was rapidly expanding from its pre industrial origins.

A different picture emerges in Bradford Dale. The transition to industrial scale mining had begun much earlier due to the demands of the iron companies. Here there was an expansion both in colliery numbers and an increase in those capable of producing over 10,000 tons per annum. Whilst retaining the more primitive sites, the iron companies were now driving up production by sinking modern collieries using steam powered winding to raise the coal. This does not mark a complete transition in working methods since throughout the century such was the demand for coal and indeed ironstone that the two continued to operate side by side.

By 1861when the main railway routes had been established and a network of branch lines though not complete had spread throughout the District, there was an expansion of colliery numbers on the outskirts of Huddersfield, notably around Almondbury, reflecting the policy of the Ramsden estate to permit mining with few restrictions in an area where no significant building projects were being contemplated. North of Halifax there was again an expansion in numbers both in Southowram and Northowram. The picture that emerges then is of a coal mining industry which because of its ability to relocate was able to resist the penetration of competition by the railways into the local market. This mobility was to keep the industry viable despite the loss of the locational advantage which it had previously enjoyed before the arrival of the railway.

Predictably in Bradford Dale the continuing demand for coal by the iron companies also saw an increase in colliery numbers as more leaseholds on Beaumont and Savile property were exploited. These colliery numbers remained high through to the 1880s when production finally began to decline as leaseholds were worked out. This was a common feature throughout the District as the tables shown in the previous chapter revealed. Effectively the industry remained in operation supplying coal on an industrial scale for about forty years even though demand in Huddersfield and Halifax could have been met by the railways[1].

It will be evident then that one of the factors which kept the industry in operation was devised by coal masters in response to fast changing conditions but this could only have happened if the decision making process undertaken by the principal landowners allowed it to happen. The three principal options available to the landowner were briefly referred to in chapter eight. It now remains to evaluate these options in greater detail. This forms the basis of using Game Theory as an analytical tool.

Prior to the 1840s there is little evidence of the Ramsden estate prioritising any particular activity. Indeed the construction of a cloth hall was the only physical sign of the estate doing anything tangible to promote the major non agricultural activity on the property. Even this investment was undertaken at a point where the trading of cloth by individual weavers was in decline. The first indication of strategic planning for land use was then the bold act of refusing to issue new mining leases in the central area of the town. This was at a time when the only alternative source of coal was via supplies brought in by canal. With the likelihood of seasonal interruptions to traffic along the canal it was not a source of supply that could always be relied upon. Coal masters were anxious to secure leaseholds in the central area of the town as this was becoming the focal point for industrial and urban expansion. Allowing mining in this area would arguably

---

[1] Hunt R. 'Mineral Statistics'

have enhanced the locational advantages of local coal, bearing in mind the limited market area of small collieries and the obvious difficulty of transporting a bulk commodity any distance.

The Ramsden estate could then have permitted mining and benefited from the income. There was of course the risk of damage although a standard clause in mining leases required the lessor to make good any surface damage. This was easier said than done. The estate had in the past been affected by subsidence due to the presence of abandoned mine workings[2] and these were small ventures with very little capital behind them. It is unlikely that a small coal master irrespective of the terms and conditions in the lease he had signed, would have been in a position to carry out remedial work. The risk factor for the estate given the by now growing value of the land in the central area and the number of building projects being proposed was too great[3]. By the 1840s the estate would also have been aware of the proposal to construct a trans Pennine Manchester to Leeds railway. On its arrival in Huddersfield this would have the effect of solving the coal supply problem and rendering the question redundant. The solution chosen by the estate appears as the classic minimax theorem; that there is always a rational solution to a precisely defined conflict between two parties whose interests are completely opposite. It is a rational solution in that both parties can convince themselves that they cannot expect to do any better, given the nature of the conflict.

This decision created a secondary risk for the Ramsden estate. Having concluded that reserving the central area for building projects was the best option for maximising income from the land, there was the possibility that demand for coal would outstrip supply. To some extent this happened and appears to have been a risk the estate was prepared to run. The wool textiles industry of the early nineteenth century was still heavily dependant on water

---

[2] Ramsden Estate Papers RA4
[3] Ramsden Estate Papers RA4

power which meant that mills were concentrated in valley bottom sites. Why the industry was slower to innovate than the cotton industry has never been explained[4]. However it is suggested that in what was to become one of the principal wool textiles towns in Yorkshire, the lag behind the Lancashire cotton industry in adopting steam power was due to the block on development of the local coal trade imposed by the Ramsden estate.

For the coal master, bearing in mind that relocation was a relatively easy and low cost option, there were other areas which might have been exploited beyond the District. However the low cost of a leasehold in Huddersfield, sometimes as little as £40 per acre, would have been an inducement to remain. Ultimately the option chosen was to locate on leaseholds belonging to the Ramsden estate on the Almondbury property. This explains the colliery location pattern. Given the destructive nature of primitive mining and the surface damage that it often brought, this option had an additional attraction for the small coal master who was unlikely to have to bear the costs associated with damage. Additionally there was little likelihood of conflict with other forms of land use on what remained a largely low grade agricultural property. Further, as the urban and industrial area of Huddersfield began to expand, collieries on the Almondbury property were able to find a growing market for the coal produced. The dense concentration of collieries on the southern outskirts of the town also links to the growth of industry extending eastwards along the River Colne.

The Ramsden estate policy was then a calculated risk. By reserving the central area for building development until the arrival of the railway in 1846 it reduced the locational advantage of local coal. This was seen as acceptable, since the mobility of small collieries meant that there was sufficient flexibility. The coal masters were

---

[4] Dinsdale A. 'Yorkshire Mill Town: a study of the spatial patterns and processes of urban industrial growth in Halifax 1801-1901 ' unpubl. PhD. Thesis Univ. of Leeds (1974)

able to exploit leaseholds in less advantageous locations because the Ramsden estate was willing to allow mining to be the principal form of land use on the periphery as the town expanded.

The Ramsden estate possessed the strategic advantage of occupying all of the central area of Huddersfield which brought with it a greater control over the decision making process. This had the effect of minimising the impact of other variables, such as the conflicting demands on land use. As a result the decision making process was much easier and the exercise of the minimax theorum rather less complicated. Smaller landowners were in a less favourable position. This was particularly the case for the Thornhill estate whose property was situated on the north east side of the town. It will be recalled that the estate took the decision to directly involve itself in the development of the coal reserves. The Thornhill property was adjacent to land which had by the 1840s become the focus of considerable mining activity. The estate took the decision to invest in a mining venture capable of producing more than 10,000 tons per annum. This direct involvement in coal mining was based on the premise that the industrial expansion of Huddersfield would extend out towards their property. In doing so the estate failed to take into account the fact that this elevated location above the industrial grime of Huddersfield might lend itself to a different form of exploitation. The result was that the neighbouring Ramsden property was to prove more suitable for upmarket housing developments. The Thornhill estate had then decided to invest in the kind of mining operation which would for example have suited property that was within the market area of the Bradford iron companies. The only way out of this impasse when it became obvious that such operations were denied a local market was to approach the Ramsden estate with a view to constructing a mineral railway which would have extended the estate's market area and linked the collieries with the expanding industrial area in the Colne Valley. Mention has already been made of the fact that this was not permitted. Again it is possible to see that the Ramsden estate, indirectly was reinforcing its policy by not permitting even mining *related* developments on that part of the estate. The Thornhill collieries had then been sunk on the premise that a market would evolve in close proximity making the capacity of more than 10,000 tons per annum a viable investment. The Thornhill estate by becoming directly involved in mining was in no position to bargain. The Ramsden estate saw no benefit in permitting the construction of a mineral railway and as a consequence the Thornhill estate had no room for manoeuvre. This illustrates a potential hazard of direct involvement in a mining venture by an estate which based its decision making on a speculative prospect. Its bargaining position with the Ramsden estate was non existent and it failed to consider if any alternative options were available and indeed appears to have ignored the fact that the majority of mining developments in the area were of a more primitive kind and therefore possessed the necessary flexibility to adjust to changing market conditions. In game theoretic terms the Thornhill estate did not start out with a 'least worst' option.

The adjacent estate to the Thornhill property was that owned by the Radcliffe baronets. This was a more isolated property with different operating conditions. For example it lay some distance from the expanding urban area. There was little prospect that the property would become attractive for either urban or industrial developments to any great extent. For the estate the factors to be taken into consideration were the thin coal seams and the prospect of railway competition eroding the limited market opportunities. For the coal master the estate was not a particularly attractive prospect for acquiring a leasehold. It was usually the case for an estate which was not directly involved in the promotion of mining ventures to simply agree terms and then show no further interest in the operation.

The Radcliffe estate was though in a weaker bargaining position than most landowners and as a consequence it seems to have appreciated the need to offer additional inducements to coal masters. Possibly the estate may have been convinced that it was operating against a time imperative, caused by the possibility that the railway might eventually remove what limited market opportunities existed. The concessions that the estate made were to grant shorter leases, and also invest in infrastructure, notably roads suitable for moving coal. It was usually the case that a land owner would expect to capitalise on wayleave by permitting the coal master to construct an access road and then make a charge for the coals carried on it, usually one halfpenny per ton. Clearly in order to expedite mining operations the estate was prepared to settle for less. Additionally by granting shorter leases the estate was increasing the prospect of wasteful mining methods and surface damage, which subsequently proved to be the case. The benefit to the estate was that it secured mining revenue albeit somewhat reduced. This makes an interesting contrast with the Thornhill estate where investment was made in a more sophisticated mining venture which was never fully realised. A game theoretic explanation shows that settling for less aided the Radcliffe estate. This was a situation of 'imperfect information'. The Radcliffe estate appreciated the threat of competition from the railway but did not know if or when it might come. Nevertheless it adopted a strategy to cater for the threat. Similarly the Thornhill estate also had imperfect information but failed to take this into consideration. It could not be sure that the market it anticipated would actually arrive and of course it failed to do so. Despite this the estate went ahead with an investment which was to fail

The Lister estate is an example of a property where revenue from agriculture was never significant. The estate was strategically situated on the edge of an expanding industrial town. Clearly the land would have to be exploited and the 'do nothing' option was of no appeal. The estate could have noted the expansion that was taking place and perhaps left the property untouched until urban development spread and the value of the land increased as demand rose. This was not an option that the estate chose. Instead by exploiting the mineral reserves on the property directly the estate retained full control. In doing so the estate incurred all the risks. The coal reserves were substantial since the property extended to 400 acres. The thickness of the main seam was 2ft with lesser though still useful seams to be found together with fireclay. The choice of direct involvement; effectively reinvesting profits from mining in order to upgrade existing collieries and open new sites retained both control and risk within the confines of the estate. The chief risk was the loss of markets as the railways arrived in Halifax. It will be appreciated that the Shibden Hall collieries lacked the mobility of more primitive undertakings operated by independent coal masters who could abandon a site when the market disappeared and relocate. Like the Thornhill estate investment in mining was literally and figuratively a sunk cost. In the case of the Thornhill estate the failure of the anticipated market to develop caused the collieries to under perform. The Lister estate avoided this experience by moving gradually to more sophisticated operations and was able to justify this investment by securing at least one major contract. It continued to have a presence in the local market for nearly forty years until the reserves began to decline and there was little opportunity to secure leaseholds on neighbouring properties. By assuming the entire risk the Lister estate was adopting the winning strategy although there was effectively no fall back position. Some risk was involved because in common with most other estates there was little recourse to professional advice. Essentially the estate sought to buttress its position as a significant supplier to local industry by increasing output in order to compete with growing competition from the railways.

Each of the two Bradford iron companies had a strategic relationship with a major landowner. It now remains to explore what was the most productive unwritten partnership between landowner and consumer. It will already be apparent that such was the dominant position of the iron companies in Bradford Dale that an extractive industry would inevitably be the major form of land use. Indeed this had begun to be the case in the eighteenth century and was already well established by the 1840s prior to the arrival of the railways. The combination of coal and iron ore occurring in association meant that even after the arrival of the railway the local market was not affected; the coal and iron ore being economically interdependent.

For the Beaumont and Savile estates cooperation with the respective iron company was clearly in the best interests of the landowners. Admittedly such was the market for coal throughout greater Bradford that it is unlikely the estates would have had difficulty finding other lessors prepared to work the coal. However the disadvantage of this approach would have been that the ironstone could not be so easily sold. Further, the iron companies were also prepared to mine and dispose of inferior coal types mined from thin seams that might themselves have been uneconomic. There were able to do this (and of course pay royalties) partly because of the creation of a wagon way network which covered a considerable area and also because they could dispose of the coal in the local domestic market; one which had grown up because of the presence of the iron companies.

This strategic relationship was not entirely harmonious. It will be recalled that the iron companies were extracting coal from their own collieries and also using more primitive sites operated by pit takers. This meant that surface damage could be considerable; wasteful methods of working were common and encroachments across leasehold boundaries also occurred. Indeed such was the need to extract coal and iron ore that these widespread mining operations often operated in a haphazard and unsystematic fashion. There was a compromise to be arrived at here: maximum effective land use/ greatest income/ least damage would have been the preferred solution for the estates. Apart from mining the infrastructure associated with operations together with the space needed to dispose of waste, meant that this would have been a difficult solution to achieve. The estates could have attempted to lease to others but the advantage of the iron companies were their size and wealth. In effect they were easier to deal with. Even the pit taker that most unreliable of lessee because of his limited financial resources, was governed by the iron company and this provided a measure of protection for the landowner. Using game theoretic analysis of the decision making process shows that the landowner achieved the greatest possible benefit for providing unrestricted access to the minerals the iron companies needed. In return the companies paid handsomely for this mineral largesse. In game theoretic terms this is the Nash Equilibrium in action: each player selects a strategy and every player gets a pay off.

These strategic partnerships went even further with the estates actively constraining the freedom of neighbouring smaller landowners to seek whatever partnerships they might have desired, by refusing to grant wayleave for the transportation of coal across their properties. The benefit for the iron companies then extended beyond the Beaumont and Savile estates to a complete domination of the mineral wealth of Bradford Dale. This was an arrangement that continued until the reserves began to decline and the companies were by the 1880s compelled to secure more of their supplies beyond Bradford and brought in by railway.

Each landowner might then be said to be playing a 'symmetric game'. They were all in the same position, that is to say each possessed low grade agricultural property beneath which lay mineral deposits, chiefly coal. Each was in proximity to an expanding industrial town or city with a rising demand for coal. Finally all were conscious of the looming threat from railway competition. The landowners set out to make rational choices based on the available information. One area where most lacked sufficient information was the geological factors and given this the surprising failure to seek expert advice. Even the Bradford iron companies, centres of the best technical knowledge, committed errors, allowing mining operations to encroach on property they were not leasing and only in the 1880s was a review carried out of Low Moor leaseholds on Beaumont property to assess the extent of illegal workings. The optimal strategy shared by all was to extract as much wealth from mineral deposits as possible, in some cases whilst there was still time to do this, before competition entered the market. One significant area of difficulty for the landowner was the willingness of the railway companies to make seemingly irrational choices. As was shown in chapter six the railway companies did not necessarily build a line purely for commercial reasons. Therefore this tendency to make an irrational choice may have been a factor affecting the landowner decision making process. The minimax approach, extracting maximum benefit whilst minimising damage is best illustrated by the Radcliffe estate, which permitted low level and destructive mining on the property and even made investments to facilitate this whilst a market for the coal still existed. Where the dominant strategy was shaped by the aim of mining the coal deposits then geography mattered most since the complexities of economic geology were not often fully considered.

Some strategies adopted by landowners took into account the various factors. The Ramsden estate was fortunate in its refusal to allow mining in the central area of the town in that it probably did no more than slow the adoption of stream power in some of the local mills. Only the flexibility of the coal mining industry prevented this strategy from having a negative impact which might have impeded the expansion of Huddersfield; since one of the chief locational advantages was being lost. Game Theory suggests that this was a high risk this strategy, offset by the mobility of the colliery trade and the willingness of landowners not to stick too rigidly to the terms of a lease. Had landowners insisted that small coal masters possessed the necessary financial resources to make good say, surface damage then the picture could have been somewhat different and mobility lost. Landowners were then willing to settle for less than the maximum which was: extraction of the coal and surface damage made good. This was a realistic compromise in a shallow coalfield with limited reserves.

Two themes have then been considered. Firstly how a coal industry was able to support industrial expansion and survive the arrival of competition from the railways. Secondly how landowners made strategic choices for the exploitation of their properties. With the exception of the Bradford iron companies where the association of coal with ironstone made investment in modern collieries worthwhile elsewhere primitive undertakings dominated the industry. This was the method best suited to working thin seams that might otherwise have been uneconomic. These primitive collieries requiring little in the way of investment possessed the great advantage of mobility and using this enabled the local industry to continue in operation whilst facing competition from the railways until reserves had been exhausted. Primitive mining methods developed for a domestic market were also suited to local conditions during the industrial age.

Landowners who were prepared to settle for less often made the best strategic choice. Those prepared to take risks did so successfully when other factors were on their side, notably the size of the coal reserves available, or the ability of the industry to adapt. Risk increased where there were factors beyond their control or they lacked sufficient information for coherent decision making. Decision making could be made easier when a strategic partnership was available which brought with it a guaranteed demand for the coal and evidence that this would continue because the other party was prepared to invest in the necessary infrastructure. Overarching all forms of decision making was the desire among landowners to gain an income from the coal reserves whilst the opportunity existed.

Acknowledgements

Bradford Public Library
Bretton Hall College, Beaumont Estate Archives
Brotherton Library, University of Leeds
Brown Library, Liverpool
Dewsbury Public Library
Halifax Public Library
Huddersfield Public Library
Huddersfield Tolson Museum
Huddersfield University Library
Leeds City Archives
Leeds Public Library
Library, North of England Institute of Mining & Mechanical Engineers, Newcastle
Library, Royal Geographical Society
West Yorkshire Archives
Yorkshire Archaeological Society

www.ingramcontent.com/pod-product-compliance
Lightning Source LLC
Chambersburg PA
CBHW031416210526
45464CB00005B/1904